Building Wireless Sensor Networks

Robert Faludi

O'REILLY®

Beijing · Cambridge · Farnham · Köln · Sebastopol · Tokyo

Building Wireless Sensor Networks
by Robert Faludi

Copyright © 2011 Robert Faludi. All rights reserved.
Printed in the United States of America.

Published by O'Reilly Media, Inc., 1005 Gravenstein Highway North, Sebastopol, CA 95472.

O'Reilly books may be purchased for educational, business, or sales promotional use. Online editions are also available for most titles (*http://my.safaribooksonline.com*). For more information, contact our corporate/institutional sales department: 800-998-9938 or *corporate@oreilly.com*.

Editor: Brian Jepson	**Indexer:** Angela Howard
Production Editor: Adam Zaremba	**Cover Designer:** Karen Montgomery
Copyeditor: Sharon Terdeman	**Interior Designer:** David Futato
Technical Editors: Kate Hartman and Jordan Husney	**Illustrator:** Robert Romano
Proofreader: Sada Preisch	

December 2010:   First Edition.

Revision History for the First Edition:
2010-12-13	First release
2011-01-28	Second release
2011-08-12	Third release
2012-01-13	Fourth release

See *http://oreilly.com/catalog/errata.csp?isbn=9780596807733* for release details.

ISBN: 978-0-596-80773-3

[LSI]

1326394422

Table of Contents

Preface

Building Wireless Sensor Networks is an essential guide for anyone interested in wireless communications for sensor networks, home networking, or device hacking. It is a first step in becoming proficient in making these systems. It is not a textbook on protocols or a complete guide to networking theory. No engineering or computer science background is expected or required. Those who have fooled around a bit with electronics or programming will certainly have a leg up, but in general, this book is aimed at hobbyists, students, makers, hardware hackers, designers, artists, and prototypers. In the chapters to come, you will scaffold your way up toward greater comfort and proficiency with hardware, software, radio, and communications. I'll explain everything necessary to get started, at least briefly. We'll create examples using accessible environments, such as Arduino for hardware and Processing for displays. And I'll provide a full range of resources, including helpful references to outside works for the electronics and networking novice. Whether you are a young inventor or an experienced engineer, this book focuses on getting your projects up and running as efficiently as possible.

All the projects you'll create in this book use radio signals that pass invisibly through the air. This "wirelessness" is essential whenever you want to place sensors where no cables can be installed, or where such tethering is undesirable. With radio, you can employ sensing and actuation in pristine natural settings, minimalist building interiors, or complex urban environments. Mobile devices like children's toys can benefit greatly by being communicative without being chained to the wall or to each other. Sensors can be attached to people or animals in a humane manner that doesn't hinder their movement. In short, lots of data can move freely from where it is gathered to where it can do the most good. That's why wireless is worth it.

The ZigBee protocol is a very popular way of creating radio sensor networks for a number of reasons. Wireless networks and connected devices in general tend to be used in situations where power is hard to come by and must be conserved. Many times the communications these networks send are small in nature, compared to systems that transfer huge files such as videos. Often, each device in the network transmits or receives unique information, so a robust system of individual addressing is extremely helpful. Security and design flexibility are frequently indispensable. That's why this book focuses on ZigBee, the protocol defined by various industry players who together form the ZigBee Alliance (*http://zigbee.org*). In the past few years, ZigBee has found its way into commercial systems for home automation, smart energy systems, consumer electronics, industrial sensing, and health care. It features full addressing, many power-saving options, optimizations for efficiency in low-bandwidth applications, and a layered approach to communications design and security. Most importantly, ZigBee automatically forms entire networks that can heal themselves, routing around problem areas without manual intervention. Designers, hackers, inventors, artists, and engineers are currently making use of this popular wireless protocol to create the systems that inform, enable, and delight their various users.

We will make a new project in almost every chapter of this book to demonstrate how everyday people, not just electrical engineers and computer scientists, can develop these systems. A number of full sensor networks, an array of doorbells, a two-way lighting detector, a household control system, and several types of Internet-connected contraptions will be demonstrated step by step for you to build. After reading this book you'll have a solid understanding of what it takes to create scalable sensor and device networks because you'll have brought a variety of them into being with your own hands. This book's website (*http://oreilly.com/catalog/9780596807733*) makes even more resources available to you.

You may wonder what drives humans to create reactive sensor systems and connected devices. Since before written history, there have been people and cultures that believed every object in the world was imbued with spirits—that even rocks are alive. This worldview, termed *animism* by modern scholars, isn't something validated by science. And yet the tacit belief that objects are in some way alive seems to resonate as a fundamentally human way of thinking. "That mixer doesn't like it when the batter is too thick." "The DVD player doesn't want to eject that disk." "My computer hates me!" We seem to want our things to be alive and frequently consider them willful—though, on an intellectual level we know they aren't. This book isn't about animism, of course; it's about making networks using ZigBee radios. However, one reason we do this— our motivation for making systems that are sensitive, active, reactive, and communicative—could just be some inherent desire to create the world we believe should exist: one where everything is imbued with a willful spirit and works together to help us live more richly. If so, this book is offered as a practical step in the right direction. I hope it will help you bring your own creations to life.

How This Book Is Organized

The chapters in this book are organized as follows:

Chapter 1, *Getting Ready*
> This chapter offers a shopping guide and an introduction to all the major components we'll be using. We focus on just what you need to get up and running, including XBee radios, adapters, breakout boards, terminal programs, and software.

Chapter 2, *Up and Running*

Right at the start of the book, you'll go from a bag of parts to a working ZigBee network in one chapter, taking the simplest path to early success. Radios, ZigBee, networks, and addressing are introduced, and then you'll configure your components to achieve a simple chat session.

Chapter 3, *Build a Better Doorbell*

This section focuses on creating something practical using the Arduino microcontroller system, which is briefly introduced. After getting up to speed on basic serial concepts and simple protocols, you'll execute a series of doorbell projects that increase in creative complexity as you gain skill.

Chapter 4, *Ins and Outs*

Here you'll take a closer look at the unique features of the XBee-brand ZigBee radios so we can start building fully scalable sensor networks. You'll begin with input/output concepts and commands, then immediately put these to use in a small set of progressively intricate projects.

Chapter 5, *API and a Sensor Network*

At this point you have everything you need to conquer the XBee's application programming interface. We start by laying a foundation of ideas and scaffold you up to a full understanding of the structured API communication frames. You are then ready to create a fully scalable sensor network of your own, using the complete example at the end of this chapter.

Chapter 6, *Sleeping, Then Changing the World*

We ease the development pace a bit here to address some nuances of ZigBee mesh networking, including sleep mode, end devices, and power management. Then it's time to change things in the physical world using direct actuation. This chapter features a powerful control project you can use to automate your home or turn just about anything on and off remotely.

Chapter 7, *Over the Borders*

In this chapter you learn to make gateways that connect ZigBee with neighboring networks, including a remarkably easy path to the Internet. You'll see full examples, showing how to allow anything to talk to everything everywhere—plus there's a special project for starry-eyed celebrity fans.

Chapter 8, *More to Love*

The final chapter is really a broader introduction. We explore advanced ZigBee techniques, demonstrate how to publish and share your data online, and then wrap things up with a peek at where ZigBee is headed.

Appendix, *Resource Guide*

To ensure that the book remains useful even after you have read it, we've included links to online resources and other texts for learning more about Arduino, Processing, Python, and ZigBee, along with a handy troubleshooting guide to get you unstuck from common mistakes. There are also tables to use as a fast daily reference to Digi radios, other brands of ZigBee modules, network analyzers, packet sniffers, connectors, shields, hexadecimals, binary numbers, ASCII codes, and finally a complete guide to XBee AT commands.

About the Title

You will notice that for a book called *Building Wireless Sensor Networks*, we spend quite a bit of time talking about *actuation*: outputs that make things happen in the physical world. The source of this is a deep-seated point of view that is backed up by some long-standing cognitive science.

"Thinking is for doing" is a phrase popularized by social psychologist Susan Fiske. Her point (and William James' when he commented similarly a century earlier) is that our brains exist first and last for creating physical actions. In fact, the brain is just the midpoint of the perception-action chain. The real action starts with our sensory systems. We see, smell, and feel, then we process those sensations for the purpose of choosing and executing our next move. Sensing never happens in a vacuum for its own sake. There's always a physical purpose. This is as true for wireless networks as it is for living organisms. The data we collect is always aimed at an action of some kind. Alarm systems trigger an immediate police response, while environmental sensing studies often have a much longer cycle that results in policies to guide real-world development. In both cases there's a purpose to the sensing that ends up, sooner or later, creating changes in the physical world. This book takes a comprehensive approach to cover both the input and output stages of the information-action cycle—sensing *and* actuation. In doing so, we hope to encourage projects to do more with data than simply collect it, hopefully enabling implementations that use their sensory input to create the rich physical experiences that humans crave.

About the Examples

All of the example circuits and code in this book are designed with clarity in mind. Astute electrical engineers will certainly notice that some corners have been cut. For example, we draw close to the rated output for the 3.3-volt pin on the Arduino board in some projects, and we rely on the microcontroller to throttle the current going to LEDs where we can. While that wouldn't be advisable in a commercial product, it does produce working circuits that are very simple for the beginner to build and understand. The same is true for the example code. Production-quality programming usually includes much more error correction and thriftier memory management than we offer here. Our purpose is to strip the code down to the basics as much as possible so that it can serve as a transparent tool for learning.

If you prefer to enhance the circuits and code to make them more robust, by all means do so! Feel free to share your suggestions or enhancements on the forums, and by sending them to us at *bookquestions@oreilly.com*. Feedback and community participation is always welcome!

Additional code and circuit diagrams that are made available in the future will be linked from this book's website (*http://oreilly.com/catalog/9780596807733*).

Conventions Used in This Book

The following typographical conventions are used in this book:

Italic

> Indicates new terms, URLs, email addresses, filenames, and file extensions

`Constant width`

> Used for program listings, as well as within paragraphs to refer to program elements such as variable or function names, databases, data types, environment variables, statements, and keywords

`Constant width bold`

> Shows commands or other text that should be typed literally by the user

`Constant width italic`

> Shows text that should be replaced with user-supplied values or by values determined by context

 This icon signifies a tip, suggestion, or general note.

 This icon signifies a warning or caution.

Using Code Examples

This book is here to help you get your job done. In general, you may use the code in this book in your programs and documentation, and the projects as a foundation for creations of your own. You do not need to contact us for permission unless you're reproducing a significant portion of the code or schematics. For example, writing a program that uses several chunks of code from this book does not require permission. Selling or distributing a CD-ROM of examples from O'Reilly books does require permission. Answering a question by citing this book and quoting example code does not require permission. Incorporating a significant amount of example code from this book into your product's documentation does require permission.

We appreciate, but do not require, attribution. An attribution usually includes the title, author, publisher, and ISBN. For example: "*Building Wireless Sensor Networks* by Robert Faludi. Copyright 2011 Robert Faludi, 978-0-596-80773-3."

If you feel your use of code examples falls outside fair use or the permission given here, feel free to contact us at *permissions@oreilly.com*.

Safari® Books Online

Safari Books Online is an on-demand digital library that lets you easily search over 7,500 technology and creative reference books and videos to find the answers you need quickly.

With a subscription, you can read any page and watch any video from our library online. Read books on your cell phone and mobile devices. Access new titles before they are available for print, and get exclusive access to manuscripts in development and post feedback for the authors. Copy and paste code samples, organize your favorites, download chapters, bookmark key sections, create notes, print out pages, and benefit from tons of other time-saving features.

O'Reilly Media has uploaded this book to the Safari Books Online service. To have full digital access to this book and others on similar topics from O'Reilly and other publishers, sign up for free at *http://my.safaribooksonline.com*.

How to Contact Us

Please address comments and questions concerning this book to the publisher:

O'Reilly Media, Inc.
1005 Gravenstein Highway North
Sebastopol, CA 95472
800-998-9938 (in the United States or Canada)
707-829-0515 (international or local)
707-829-0104 (fax)

We have a web page for this book, where we list errata, examples, and any additional information. You can access this page at:

http://oreilly.com/catalog/9780596807733

To comment or ask technical questions about this book, send email to:

bookquestions@oreilly.com

For more information about our books, conferences, Resource Centers, and the O'Reilly Network, see our website at:

http://oreilly.com

Acknowledgments

This book was strongly affected by my tag team of editors Brian Jepson and Tom Sgouros. Brian's fractured wit paired with his expert strategies constantly guided my hand, while Tom's attention to details and scientific discipline provided the rigor any technical book demands. Even when process and schedule left me breathless, I never lost appreciation for the wisdom and craft they supplied. I'm grateful for all their help.

My technical editors imparted a level of feedback that went well beyond their respective calls of duty. Kate Hartman, who encouraged this book from the get-go, spent many hours combing the text for confusing constructions and omitted explanations. Her project assessments and clarity of voice are felt throughout. Jordan Husney cheerfully reviewed many of these chapters from his perch at 35,000 feet. His deep knowledge of the ZigBee protocol is matched only by his competence as a wordsmith. Thanks to both for their efforts and uncommon friendships.

Building Wireless Sensor Networks is loosely structured around the Sociable Objects class I created at NYU's ITP graduate program in media and technology. There, Tom Igoe loaned me my first ZigBee radio, encouraged my projects, mentored my development as a teacher, and continues to be a seemingly bottomless well of excellent advice and terrible puns. This book almost certainly could not have happened without him. Clay Shirky, Nancy Hechinger, Marianne Petit, Dan Shiffman, Danny Rozen, and Dan O'Sullivan are but a few of the instructors who provided invaluable inspiration. George Agudow and the sensational staff at ITP have granted support to my work at every turn. My fellow resident researchers Jeff, John, Jenny, Kate, Gabe, and Demetrie influenced my ideas and enriched my experience during the year we were all lucky enough to work together. Almost all the concepts in this book were trialed by my Sociable Objects students and I am grateful for their feedback, which is incorporated throughout. Everyone in the ITP community owes a debt toward longtime Chair and perpetual guiding star Red Burns. Her steely stare, firm love, and rare brilliance continue to illuminate us all.

This book would have been immeasurably more difficult without Paul Cole's flexibility, generous spirit, and unflagging support. I am thankful for the grand company of my entire day job crew at GroundedPower, especially longtime collaborators Terence Arjo, Mike Bukhin, and Demetrie Tyler. They caught my bullets on countless occasions when I needed extra concentration for penning these pages. At SVA's MFA program in Interaction Design, Liz Danzico's words of wisdom and my graduate students' insightful creations brought depth to my thinking and clarity to my explanations.

My mother and father taught me to craft with words, wood, and wires—priceless skills that I am honored to share in some small measure here. I am lucky to come from two people with such talent, creative motivation, and quick-witted humor. My sister, Susan, and her partner, Russ, tirelessly guided me through the tricks of the writing trade. When enthusiasm flagged, Sue and Russ assured me that my writerly doldrums were distinctly underwhelming, cannily undermining my laments and restoring my cheer. I'm phenomenally lucky to have them in my life. Liz Arum bestowed suggestions, solace, affection, and perpetual patience as I alternately plodded and sprinted through the birthing of this book. Her family has pampered me with their hospitality, and her middle school students effortlessly completed several of the book's projects, just as she knew they could. I'm grateful to all of them, and to Liz especially.

One more thing: John Dobson's telescope-building class and indomitable spirit continue to be an inspiration in my work. If you ever get a chance to build your own sidewalk telescope, don't hesitate for a minute. Everyone deserves to meet the universe in person.

 The connection diagrams in this book were created with Fritzing, an open source tool for documenting, sharing, teaching, and designing interactive electronic projects. For more information, see their website (*http://fritzing.org/*).

CHAPTER 1

Getting Ready

Let's get right down to business. This chapter offers a shopping guide and an introduction to all the major components you'll need to prep your networking toolbox with essential parts and programs. There are plenty of options, so we're going to focus on just what you need to get up and running. Check the Appendix for a comprehensive list of resources. For now, here are the essentials, distilled for your convenience.

 In this book we focus on XBee brand ZigBee radios because they have a host of features that make them especially easy for beginners to work with. Many other brands exist, however most are best suited to those with an electrical engineering background. If that's you, the resource guide at the end of this book lists other ZigBee options you can consider. Professional engineers often prefer XBees for prototyping or anytime a reduced development effort makes them the most cost-effective option.

Buying an XBee Radio

Digi International manufactures a bewildering array of XBee-branded radios. All told there are at least 30 different combinations of component hardware, firmware protocols, transmission powers, and antenna options. We'll first take a look at what's out there, and then narrow that down to the devices we will be using in this book.

Hardware

There are two basic varieties of XBee radio physical hardware:

XBee Series 1 hardware
These radios use a microchip made by Freescale to provide simple, standards-based point-to-point communications, as well as a proprietary implementation of mesh networking. We won't use the Series 1 hardware at all in this book.

The sidebar "Series 1 Radios" on page 3 takes a quick look at the Series 1, but remember that the examples in this book *won't* work with Series 1 hardware.

XBee Series 2 hardware

The Series 2 uses a microchip from Ember Networks that enables several different flavors of standards-based ZigBee mesh networking. Mesh networking is the heart of creating robust sensor networks, the systems that can generate immensely rich data sets or support intricate human-scale interactions. Everything we do in this book from here on out will use the Series 2 hardware exclusively.

 Digi has just released the newer Series 2B. Series 2B features include reduced power consumption, additional antenna options, and an optional programmable microprocessor. For the most part, Series 2 and 2B are interchangeable.

Both the Series 1 and Series 2 radios are available in two different transmission powers, regular and PRO (see Figure 1-1). The regular version is called simply an XBee. The XBee-PRO radio has more power and is larger and more expensive.

Figure 1-1. XBee radios in regular and PRO flavors

The regular version is a slightly less expensive way to get started. For now, we won't worry about distinguishing between the regular and PRO radios because they are configured in the same way.

Series 1 Radios

Series 1 XBee modules are quite popular with the do-it-yourself crowd, while Series 2 hardware supports the full ZigBee protocol. Series 1 is great for simple cable replacements and smaller-sized systems. Series 2 is designed with larger sensor networks in mind and is essential for the robust interactions with the ZigBee standards-based systems that are now being widely deployed in residential, academic, and commercial settings.

The Series 2 hardware has a little better range and uses slightly less power than the Series 1; yet these small improvements would not be a reason to choose one format over the other. Both use the same physical footprint and can be easily interchanged, often with only minor changes to the underlying software. However, the Series 2 will not talk to or interoperate with the Series 1 *at all*. Each network must use only one version. Table 1-1 shows a summary of the similarities and differences.

Table 1-1. Series 1 versus Series 2 for regular XBees

	Series 1	Series 2
Typical (indoor/urban) range	30 meters	40 meters
Best (line of sight) range	100 meters	120 meters
Transmit/Receive current	45/50 mA	40/40 mA
Firmware (typical)	802.15.4 point-to-point	ZB ZigBee mesh
Digital input/output pins	8 (plus 1 input-only)	11
Analog input pins	7	4
Analog (PWM) output pins	2	None
Low power, low bandwidth, low cost, addressable, standardized, small, popular	Yes	Yes
Interoperable mesh routing, ad hoc network creation, self-healing networks	No	Yes
Point-to-point, star topologies	Yes	Yes
Mesh, cluster tree topologies	No	Yes
Single firmware for all modes	Yes	No
Requires coordinator node	No	Yes
Point-to-point configuration	Simple	More involved
Standards-based networking	Yes	Yes
Standards-based applications	No	Yes
Underlying chipset	Freescale	Ember
Firmware available	802.15.4 (IEEE standard), DigiMesh (proprietary)	ZB (ZigBee 2007), ZNet 2.5 (obsolete)
Up-to-date and actively supported	Yes	Yes

While this book uses the Series 2 hardware exclusively, what you learn here can help you with both series. Picking up the Series 1 commands should be a snap after reading this book because, for the most part, they are a subset of the Series 2 that we cover here. You will already know pretty much everything you need to work with them! Tom Igoe's excellent volume *Making Things Talk* (*http://oreilly.com/catalog/0636920010920/*) (O'Reilly) has several appealing example projects for Series 1 XBees, and many more are available online.

Antennas

Radios need antennas to transmit and receive signals. There's more than one way to build an antenna, each with advantages and disadvantages. You probably won't be surprised to learn that Digi decided to offer plenty of choices. Here are the kinds of antenna options currently available (see Figure 1-2):

Whip or wire antenna
> This is just what it sounds like—a single piece of wire sticking up from the body of the radio. In most cases, the wire antenna is just what you need. It's simple and offers omnidirectional radiation, meaning the maximum transmission distance is pretty much the same in all directions when its wire is straight and perpendicular to the module.

Chip antenna
> Again, this is pretty much what it sounds like. The chip antenna is a flat ceramic chip that's flush with the body of the XBee. That makes it smaller and sturdier, but those advantages come at a price. Chip antennas have a cardioid (heart-shaped) radiation pattern, meaning that the signal is attenuated in many directions. However, if you're making a device where mechanical stress to the wire antenna might break it, or you need to put the radio in a very small space, then the chip antenna may be your best bet. Chip antennas are often the right choice for anything wearable.

PCB antenna
> Introduced with the XBee-PRO S2B, the PCB antenna is printed directly on the circuit board of the XBee. It is composed of a series of conducting traces laid out in a fractal pattern. The PCB antenna offers many of the same advantages (and disadvantages) as the chip antenna with a much lower cost to manufacture.

U.FL connector
> This is the smaller of the two types of external antenna connectors. More often than not, an external antenna is not needed: it is an additional expense if a simple wire antenna will do. However, when your radio is going to live on the inside of a metal box then the antenna will need to live on the outside. That way the signal is not attenuated by the enclosure. Also, it is sometimes advantageous to orient an external antenna differently than the XBee itself to or use a special-purpose antenna with a specific radiation pattern, such as a high-gain antenna that passes signals in

a single direction over a broader distance. The U.FL connector is small, somewhat fragile, and almost always used with a short connecting cable that carries the signal from a remotely mounted antenna.

RPSMA connector

The RPSMA connector is just a different type of socket from the U.FL connector. It's bigger and bulkier, but you can use it with an external antenna, mounted directly to the XBee without a connecting cable. For most introductory projects, you're still best off with the simple wire antenna that is smaller, cheaper, and usually just as good.

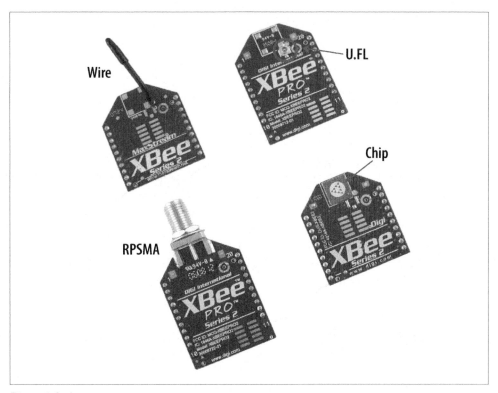

Figure 1-2. Antenna types

To keep it simple you can get started by purchasing two (or more) of the regular Series 2 XBees with wire antennas. Don't buy just one! You'll be as disappointed as a kid at Christmas who finds a single walkie-talkie under the tree. Here's the model number to get you started: XB24-Z7WIT-004. At the time of this writing, this module cost about $25.

If you need a chip antenna, the part number is XB24-Z7CIT-004. If you require a PRO high-power radio, use part number XBP24BZ7WIT-004.

Vendors

Now that you know what you want to buy, you also need to decide where to buy it. XBee radios are available directly from Digi, and also from many online resellers. This is a list of vendors that typically stock XBees, as well as many other nifty electronic components you may need for your projects:

Digi International (http://www.digi.com)
Digi manufactures and sells all varieties of XBee radios and some interesting XBee starter kits, generally at the suggested retail price. They don't sell any other electronic components.

Maker SHED (http://www.makershed.com)
MAKE: magazine (which is published by O'Reilly, the publisher of this book) offers a kit specifically designed for this book, via their in-house maker emporium. The kit includes many of the parts you'll need, including appropriate XBees.

SparkFun Electronics (http://sparkfun.com)
SparkFun carries a rapidly growing array of prototyping supplies designed specifically for DIY electronics enthusiasts, including most of the XBee modules. You'll find documentation links for each part, as well as handy tutorials for using many of the components.

DigiKey (http://www.digikey.com)
DigiKey offers a dizzying array of electronic components for the professional electrical-engineering market. It's normal to feel overwhelmed at first by the selection of about half a million different parts, but it's worth learning the ropes because you'll be able to buy almost anything you want and receive it the next day. The entire XBee line is usually represented at DigiKey (which has no relationship at all to Digi International).

Part numbers have been supplied for most of the parts recommended in this book. You'll see the vendors abbreviated in those lists: SparkFun is SFE; DigiKey, DK; Maker Shed, MS; Radio Shack, RS; Adafruit, AF.

Buying an Adapter

You'll be using a computer to configure your XBee and to send and receive data directly from your desktop or laptop. The XBee is made to be soldered directly into a printed circuit board, so you'll need an adapter to connect it to your computer's USB port. If you need to connect to an older 9-pin or 25-pin serial port instead, check the Appendix for other options.

There are several different adapters available, along with a few handy hacks if you want to avoid buying one or if you get caught without one.

Digi Evaluation Board

If you buy a complete drop-in networking starter kit from Digi, such as their *iDigi Professional Development Kit ZB* (part no. XK-Z11-PD), it will include an evaluation board with a power supply, a USB connector, and some handy buttons and lights (Digi part no. XBIB-U). The kits are a good value if you need everything they include. However, if you only want some radios and an adapter, other approaches are more cost-effective. Also, the Digi evaluation board is substantially larger than most third-party adapters, making it somewhat cumbersome to carry around. At the time of this writing, the development kit was available for $300, though occasional promotions have brought it down to $150. (See Figure 1-3.)

Figure 1-3. Digi evaluation board

USB Adapters

Several different XBee USB adapters are available from third-party manufacturers (see Figure 1-4):

Figure 1-4. XBee adapters are available from many vendors in a variety of different formats

 Almost all XBee USB adapters require drivers from FTDI (*http://www .ftdichip.com/Drivers/VCP.htm*). Be sure to install these before using your adapter.

SparkFun XBee Explorer

The Explorer is a very popular adapter that uses a fairly standard USB A to mini-B cable to connect with your computer. We'll be using it in most of the examples in this book. The cable is sold separately, but before you buy, check to see if you already have one. Many digital cameras come with this type of cable. Be aware that if you add male headers to use it in a breadboard, the pin order will not be the same as on the XBee. Check the data sheet carefully if you are using the Explorer with a breadboard setup. (About $25; *http://www.sparkfun.com/commerce/product_info .php?products_id=8687*.)

Adafruit XBee Adapter Kit

This is an inexpensive board that you'll need to solder together yourself. It also must be used with a special USB cable called the FTDI USB TTL-232, which can attach to its pin headers. The cable can be used with certain Arduino-type boards as well. Male headers can be added so that this adapter can be used in a breadboard. (About $10; *http://www.adafruit.com/index.php?main_page=product_info&cPath =29&products_id=126*. Cable about $20; *http://www.adafruit.com/index.php ?main_page=product_info&cPath=29&products_id=70*.)

New Micros XBee Dongle

One of the smallest adapters, it needs no external cable. The Dongle does not provide any access to the radio beyond USB. Also, because it has no cable, its shape sometimes interferes with other cables or the computer casing. On the other hand, it's a very small all-in-one device that's easy to carry in a pocket. It's terrific for use on the go. (About $39; *http://www.newmicros.com/cgi-bin/store/order.cgi?form= prod_detail&part=USB-XBEE-DONGLE-CARRIER.*)

Gravitech XBee to USB Adapter

Like the Explorer, this is another simple adapter board that uses the USB A to mini-B cable (not included). This one also has standard breadboard pinouts. (About $28; *http://store.gravitech.us/xbtousbad.html.*)

Breadboards

Solderless breadboards (Figure 1-5) provide an easy test bed for hooking up electronic circuits without needing to make permanent connections. They consist of a plastic housing riddled with small holes. Metal clips that lurk beneath the holes in the breadboard provide a way to hold and connect components. Each metal clip is called a bus, and everything attached to the same bus is connected together electrically. Many breadboards have two power and ground buses running down each side, with shorter buses oriented to them at right angles (Figure 1-6). Components like LEDs, capacitors, radios, and microchips are placed in the shorter buses, called terminal strips. These are connected to the power buses and each other using jumper wires.

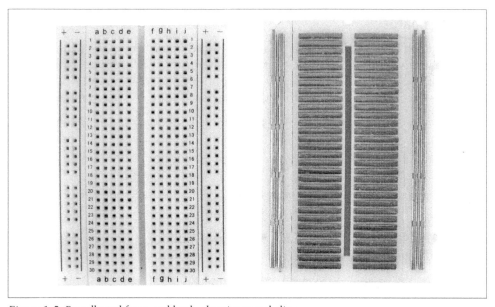

Figure 1-5. Breadboard front and back, showing metal clips

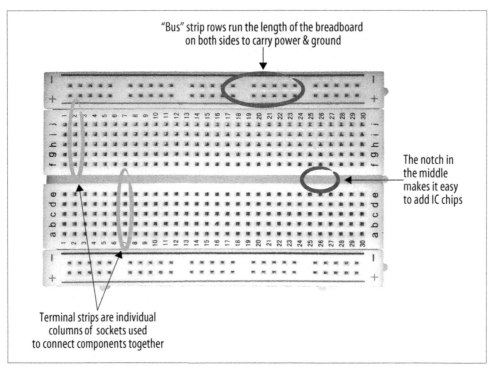

"Bus" strip rows run the length of the breadboard on both sides to carry power & ground

The notch in the middle makes it easy to add IC chips

Terminal strips are individual columns of sockets used to connect components together

Figure 1-6. Breadboard with bus strips and terminal strips indicated

Breakout Boards

All XBee radios have 20 connection pins, each spaced 2 mm apart. The tight spacing of the pins helps to keep the radios very small, but doesn't allow them to fit into a solderless breadboard. Luckily, this is a very easy problem to solve. Simple XBee break-out boards that adapt to 0.1″ breadboard spacing (see Figure 1-7) are available from:

- SparkFun (*http://www.sparkfun.com/commerce/product_info.php?products_id= 8276*)
- Adafruit (*http://www.adafruit.com/index.php?main_page=product_info&products _id=127*)
- Cutedigi (*http://www.cutedigi.com/product_info.php?products_id=4241*)
- Gravitech (*http://store.gravitech.us/xbeeadapter33v.html*)

You will generally need to solder 2 mm female pin headers to one side of these breakout boards, and regular 0.1-inch male headers to the other side.

Note that the XBee Explorer (Figure 1-8), Adafruit XBee Adapter Kit, and the MCPros XBee Simple Board each have mounting holes for 0.1-inch male headers. Solder a set of male header pins into them to adapt these for breadboard use.

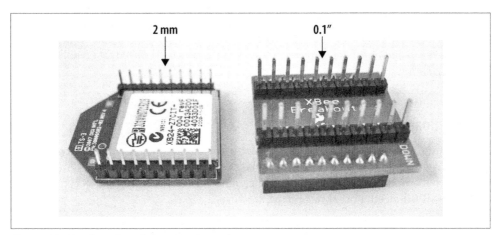

Figure 1-7. Breakout board showing pin spacing

Figure 1-8. XBee Explorer board from SparkFun

Adapters, Breakout Boards, and Shields

In case you are still a bit mystified by the different ways that an XBee radio can be attached to another device, here's a quick review:

Adapters
> Typically used to connect the XBee to a USB port on your computer. Some also provide breakout-board functionality.

Breakout boards
> Used to plug your XBee into a standard breadboard and facilitate wired connections to other components, including the Arduino board.

> *Shields*
>> These attach an XBee directly to an Arduino microcontroller. Shields are printed circuit boards engineered to seat directly on top of an Arduino board. When you are not including other components, the shield eliminates the need for breadboards and wiring.

Arduino Board Adapter Hack

The Arduino microcontroller board we'll be using in Chapter 4 can be modified to function as an adapter for XBee radios. This is a useful hack if you don't want to buy an adapter—or anytime you find yourself caught without your regular XBee adapter setup. You'll still need a breakout board, however.

 This hack allows you to connect to the XBee from a terminal program (described later in this chapter). It lets you use some features of X-CTU (also described later), but it does not support firmware upgrades. For that, you should use a proper XBee adapter.

Here's what you'll need:

- XBee radio (see "Buying an XBee Radio" on page 1)
- XBee breakout board (see "Breakout Boards" on page 10)
- Arduino microcontroller board with USB connection (Uno or similar) (SFE DEV-09950, AF 50)
- USB A-to-B-type cable (AF 62, DK 88732-9002, SFE CAB-00512)
- Solderless breadboard (AF 64, DK 438-1045-ND, SFE PRT-09567)
- Hookup wire (22 gauge or similar, different colors) or jumper wire kit (AF 153, DK 923351-ND, SFE PRT-00124)
- Wire strippers (AF 147, DK PAL70057-ND, SFE TOL-08696)
- IC extractor (DK K374-ND, RS 276-1581) or small flat-blade screwdriver (SFE TOL-09146)

 These part numbers are prefaced with abbreviations for the suppliers: DK, DigiKey; SFE, SparkFun Electronics; AF, Adafruit; RS, Radio Shack.

Insert the XBee into the breakout board, then mount the breakout board in the breadboard so that one set of legs is on each side of the breadboard's center channel. Cut four lengths of wire or select some precut jumper wires long enough to reach from the Arduino to the XBee. Use red, black, and two other colors of wire if you have them.

Figure 1-9 shows the Arduino adapter hack breadboard layout, while Figure 1-10 shows the schematic:

1. Make sure that the Arduino is unplugged from the USB and disconnected from any other external power supply before setting up your wiring.

2. Carefully remove the ATMEGA chip from the Arduino, using an integrated circuit (IC) extractor or a small flat-blade screwdriver (when you replace it later, be sure the notch at one end of the chip is aligned with the notch in the socket). Or, if you don't want to pull the chip, program the Arduino with the following null code, which prevents the Arduino chip's bootloader from responding accidentally:

   ```
   void setup() {
   }
   void loop() {
   }
   ```

3. Connect a (red) wire from the 3.3 V socket on the Arduino so that it mates with the XBee's pin 1, the 3.3 V input pin in the upper-left corner of the XBee.

4. Next, connect a (black) wire from either GND socket on the Arduino so that it mates with pin 10 on the XBee in its lower-left corner.

5. Now wire up a connection from the TX pin (pin 1) on the Arduino to pin 2, the TX/DOUT pin on the XBee. See Table 1-2 and Figures 1-11 and 1-12 for a full list of the XBee's hardware pins and their functions.

6. Finally, wire a connection from the RX pin (pin 0) on the Arduino to pin 3, the RX/DIN pin on the XBee.

7. Check all your connections. It is very important that you supply only 3.3 V power to your radio.

XBee radios will not work with voltages larger than 3.3. Giving them more than 7 volts will burn them out. When in doubt, remove the radio from your project and confirm the voltage with a multimeter (AF 71, DK BK117B-ND, SFE TOL-09141) before proceeding.

When you're done with the hack, set it aside for now. You won't need to power up this circuit until you get to "Configuring XBee" on page 40 in Chapter 2.

If you already have an Arduino Mini, you can use the same USB adapter you use for uploading code to the Arduino Mini as a connector for an XBee on a breakout board. For this adapter, wire RX to RX on the XBee and TX to TX on the XBee (*http://www.makershed.com/ProductDetails .asp?ProductCode=MKSP3*).

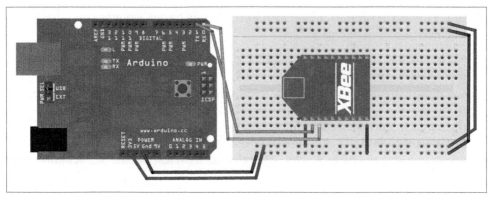

Figure 1-9. Arduino adapter hack breadboard layout

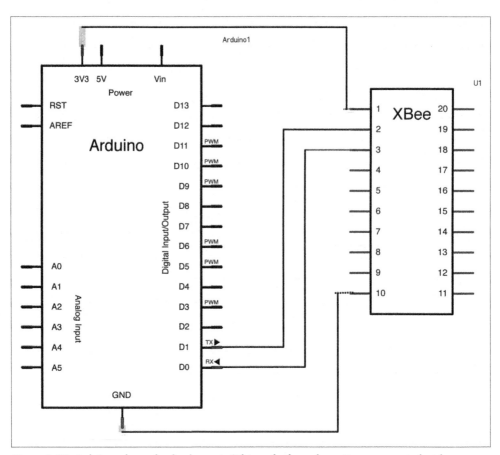

Figure 1-10. Arduino adapter hack schematic (This and other schematics were created with an open-source software package for electronic prototyping called Fritzing. Learn more at http://fritzing.org.)

What Are All Those Pins?

Table 1-2. XBee pin descriptions

Pin #	Name(s)	Description
1	VCC	3.3 V power supply
2	DOUT	Data Out (TX)
3	DIN	Data In (RX)
4	DIO12	Digital I/O 12
5	RESET	Module reset (asserted low by bringing pin to ground)
6	PWM0/RSSI/DIO10	Pulse-width modulation analog output 0, Received Signal Strength Indicator, Digital I/O 10
7	DIO11	Digital I/O 11
8	Reserved	Do not connect
9	DTR/SLEEP_RQ/ DIO8	Data Terminal Ready (hardware handshaking signal), Pin Sleep Control (asserted low), Digital I/O 8
10	GND	Ground
11	DIO4	Digital I/O 4
12	CTS/DIO7	Clear to Send (hardware handshaking), Digital I/O 7
13	ON/SLEEP	Sleep indicator (off when module is sleeping)
14	VREF	Not used in Series 2
15	ASSOC/DIO5	Association indicator: blinks if module is associated with a network, steady if not; Digital I/O 5
16	RTS/DIO6	Request to Send (hardware handshaking), Digital I/O 6
17	AD3/DIO3	Analog Input 3, Digital I/O 3
18	AD2/DIO2	Analog Input 2, Digital I/O 2
19	AD1/DIO1	Analog Input 1, Digital I/O 1
20	AD0/DIO0/COMMIS	Analog Input 0, Digital I/O 0, Commissioning Button

Choosing a Terminal Program

Each XBee radio has a tiny computer on board. This internal microcontroller runs a program, also known as *firmware*, that performs all its addressing, communication, security, and utility functions. You can configure this firmware with different *settings* that define things like its local address, which type of security is enforced, who it should send messages to, and how it should read sensors connected to its local input pins.

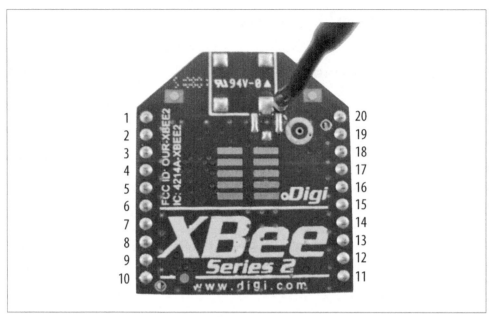

Figure 1-11. XBee physical pin numbering, front view

Figure 1-12. XBee physical pin numbering, back view

To change or upgrade the firmware, we will use a program called X-CTU that you can download from the Digi website (*http://www.digi.com/support/kbase/kbaseresultdetl.jsp ?kb=125*). On the upside, this program is totally free. On the downside, it runs only on Windows. Don't worry if you have limited access to Windows, though. Chances are you'll only need X-CTU initially, to load the proper firmware onto your XBee radio. Going forward, you can use serial terminal programs on Macintosh, Linux, or Windows to change many of the settings you'll be working with on a day-to-day basis. Let's take a look at some of these programs and how they operate.

Firmware Update Software

There is only one option for updating the low-level firmware on XBee radios: Digi's configuration tool, which is available for free.

X-CTU

The X-CTU program is the official configuration program for XBee radios. As noted, X-CTU is available only for the Microsoft Windows operating system. If you have access to a native Windows computer, a Macintosh running Windows under Boot Camp or Parallels, or a Linux computer running the WINE Windows emulator (see "X-CTU in Linux" on page 33 in Chapter 2), you're all set. Luckily X-CTU is *required* only for updating firmware, which is a relatively infrequent task. It does have a number of other handy features, though, including fully commented setup commands, range tests, and easier access to the API features we'll be examining in Chapter 5.

To use X-CTU, plug your XBee radio into a USB adapter and plug that adapter into one of your computer's USB ports. Next, launch the X-CTU program. It should show your USB connection as one of the available ports, similar to what you see in Figure 1-13. Select the appropriate port and then click on the Modem Configuration tab to get to a basic configuration screen (Figure 1-14). Clicking on the Read button will generally access the radio's setup, though this depends upon which firmware is currently loaded. Don't be concerned if you get an error message instead. We'll go over the details in the next chapter.

Terminal Software for Configuring Settings

Once you've loaded the firmware, you may want to use a different program to communicate with your XBee. It's very helpful to have some familiarity with one or more serial terminals because you may not always have access to X-CTU when you need it. At a friend's house, a hacking workshop, a public demo, or in the midst of a Maker Faire, you might need to check something or change a setting on a non-Windows computer. Or you may run into a Windows machine where you don't have the rights to download and install new software. Here's a host of different options that can save you in such cases. We'll talk about how to set them up and use them in the next chapter.

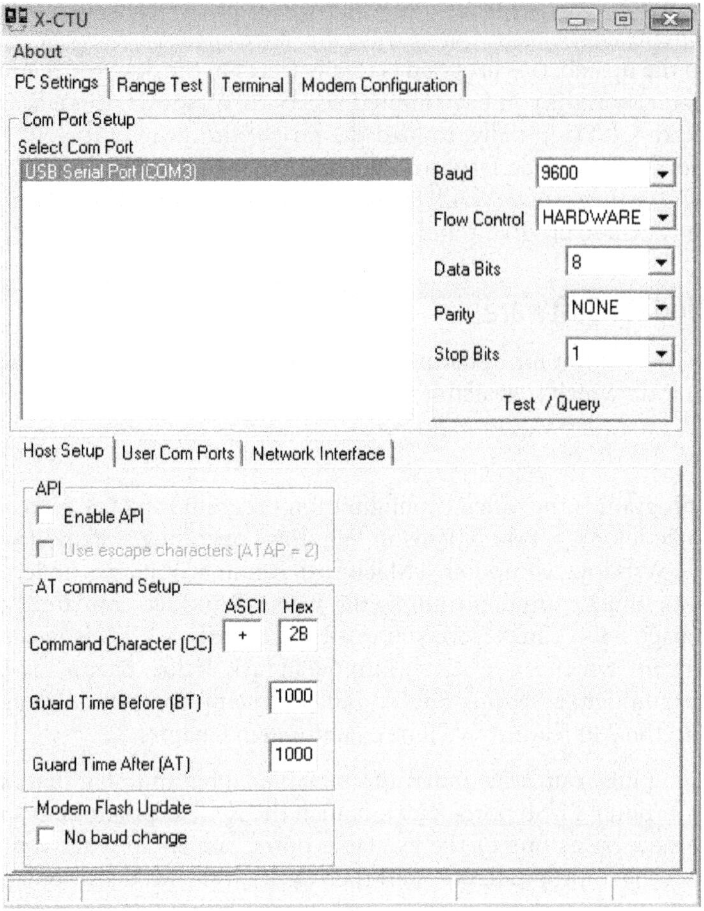

Figure 1-13. Using X-CTU

CoolTerm

CoolTerm is a terrific open source serial terminal program created by Roger Meier that runs well on both Windows and Macintosh. It's a relatively simple program that's perfect for most basic tasks you need to perform with XBee radios. CoolTerm is free. Consider making a small donation to show your appreciation and encourage continued support for the program (*http://freeware.the-meiers.org*).

HyperTerminal

Windows XP and older Windows versions come with a serial terminal program called HyperTerminal. If you are using Windows Vista or Windows 7, HyperTerminal may still be available as a free demo, or for purchase from *http://www.hilgraeve.com/hyper terminal.html*.

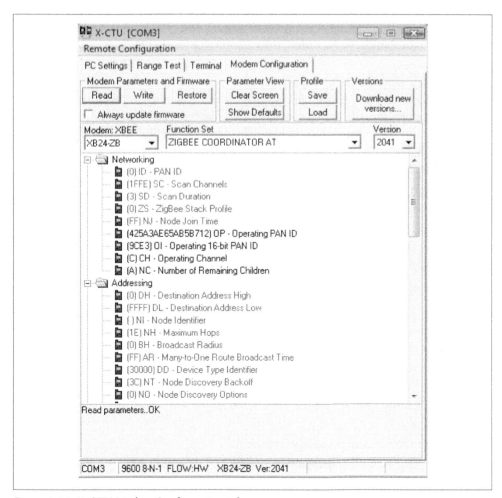

Figure 1-14. X-CTU Modem Configuration tab

Tera Term

Tera Term is a free, open source Windows program that performs a wide variety of terminal functions, including acting as a serial terminal. Those using Vista or Windows 7 will appreciate having a free option, since HyperTerminal is no longer bundled with Windows and must be purchased separately. This is the Windows software we'll use to demonstrate serial terminal use (*http://ttssh2.sourceforge.jp/*).

ZTerm

An old favorite terminal program on the Macintosh, ZTerm has been showing its age for quite some time. It was designed in 1992 and was last updated in 2002. Still, it is widely used and despite its anachronistic features and idiosyncratic design, it's been stable for almost 20 years. You'll find some brief setup documentation on my blog (*http://www.faludi.com/2009/09/25/zterm-settings/*), and you can download ZTerm and pay its small shareware fee online (*http://homepage.mac.com/dalverson/zterm/*).

screen

For Linux users and for those comfortable in the Macintosh Terminal, there's a command-line program named screen that allows direct access to serial ports, including USB devices. On Mac OS X, the command **ls dev/tty.*** will list the available ports, returning a list like this:

```
/dev/tty.Bluetooth-Modem    /dev/tty.Bluetooth-PDA-Sync /dev/tty.usb-A410032.
```

On Linux, try **ls dev/ttyUSB***. Your serial port will probably be something like /dev/ttyUSB0.

Once you know what your USB port is called, you can invoke the screen program, using the port and a data rate of 9600 baud. For example:

```
screen /dev/tty.usb-A410032 9600
```

To exit, type Ctrl-A followed by Ctrl-\ and then **y** to quit.

The picocom program, described in the sidebar "A Serial Terminal Program for Linux" on page 40 in Chapter 2, is an alternative to screen and has certain features (such as local echo) that can be useful for working with XBees.

Others

Here are some other popular options for serial terminals. Some are free, and some aren't:

- RealTerm (*http://realterm.sourceforge.net/*)
- Termite (*http://www.compuphase.com/software_termite.htm*)
- PuTTY (*http://www.chiark.greenend.org.uk/~sgtatham/putty/*)
- MacWise (*http://www.macwise.com/*)

Summary

Here is a basic shopping list that will work well for this book. Feel free to customize it according to your interests and the projects you have planned:

- Three XBee ZBs with wire antenna (Digi: XB24-Z7WIT-004, DK602-1098-ND)
- One or two XBee Explorers (SFE: WRL-08687)
- One or two USB A to Mini-B cables (SFE: CAB-00598)
- X-CTU for Windows (free)
- CoolTerm (free)

Now that you know what to get, go do it! As soon as your components arrive, you will probably be itching to use them. The next chapter will help you transform your box of parts into a working ZigBee network. You'll be chatting wirelessly in no time.

CHAPTER 2

Up and Running

Here is the heart of the book. We go from a bag of parts to a working ZigBee network in one chapter, taking the simplest path to early success. Addresses, firmware, and configuration steps culminate in a simple chat session for a satisfying exchange of greetings. Hello world, you are up and running.

Let's get started.

Radio Basics

What exactly is this thing called *radio*? In any dictionary or encyclopedia, you'll find a definition that describes the transmission of information via modulation of waves in the electromagnetic spectrum. Whoa, that's pretty mysterious, especially when coupled with the mathematics and equations that describe the behavior of radio. These certainly help us work with the medium, yet they still may not answer the question of what it *is*. If you feel unsatisfied by the words or the math, that's OK. One helpful way to think of radio is as a well-described mystery. After all, we can't see radio. We can't touch radio or hear it or smell it or feel it. Billions of years of evolution haven't provided us with any direct sensory apparatus for perceiving the radio part of the electromagnetic spectrum *at all*. Our language around the phenomenon reveals this. The word radio comes from radius, the Latin for a ray or spoke in a wheel, something that propagates from a center outward. True, but pretty vague. Around the turn of the last century it was referred to as "ethereal communication" in a nod to the "ether" that was incorrectly thought to pervade outer space. That turned out to be just wrong. Today it's often referred to as wireless communication, but that's not what it is. That's what it *isn't*. Radio is also tomato-less, cheese-less, and bread-less, but that does no better to help us understand it. The element of mystery is fundamental to the human experience of radio, and a reassuring clue that your grasp on it will always be a little loose. Luckily, we do have a tremendous ability to describe radio's propagation, predict its behavior, and use it efficiently for a huge number of purposes. For example, you are about to use it in the creation of your own sensor mesh network, cleverly manipulating a phenomenon that is beyond your direct powers of perception. That's pretty neat.

Electromagnetic Spectrum

Radio is only one slice of the broad array of energy we call the electromagnetic spectrum (see Figure 2-1). This spectrum includes high-power gamma rays that arrive from supernovas in outer space, the X-rays we use to sneak a peek at broken bones, microwaves that cook our food, infrared that we sense broadly as heat, and the one tiny slice of the spectrum that about a third of our brain is devoted to decoding: visible light. Radio waves are much longer than light waves and many can travel through opaque substances such as clothing, furniture, and brick walls. Radio energy requires no medium. It can propagate perfectly well in a vacuum and is therefore ideal for communications where metal wire connections are impractical, or where visual line of sight may be impeded by obstructions. When radio waves impinge on a conductor, like metal, they induce an electrical current that transforms their energy into another form. This means that radio will not travel well through metal walls, but it also means that we can employ metallic antennas to transduce radio energy into electrical signals that computers can detect and process. Engineers have a comprehensive body of theories, equations, and laws for predicting and manipulating the behavior of radio. Luckily, we can make do for the time being with just one of these tools, the *inverse square law*.

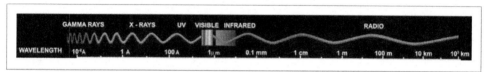

Figure 2-1. Chart of the electromagnetic spectrum

Inverse Square Law

Radio signals require a lot of power because, unlike messages running through a wire, they decay in an accelerated fashion. The reason for this is easy to understand. As radio signals radiate away from their source, they rapidly spread out like ripples in a pool. Sound works pretty much the same way, which is why it's easy to hear a whisper up close, but impossible to understand it even a few feet away. Both sound and radio decay according to the inverse square law. Each time you double the distance, you require four times the amount of power (as Figure 2-2 shows), so traversing long distances requires tremendous expenditures of energy compared to shorter ones.

ZigBee mesh networks are designed with the inverse square law in mind. Rather than using big batteries to generate the large amount of power needed to send a signal over a great distance, each radio needs only small amounts of power to go a short distance to its nearest neighbor in the network. By adding nodes to the network, great distances can be traversed without any node needing access to large amounts of energy.

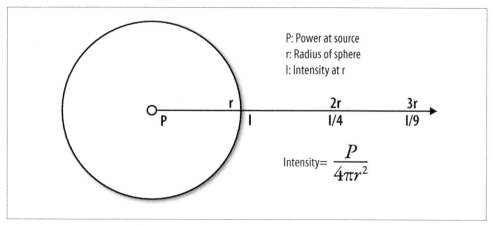

Figure 2-2. Inverse square law

Introduction to ZigBee

Many people think that ZigBee and XBee are the same thing. That's not true. ZigBee is a standard communications protocol for low-power, wireless mesh networking. XBee is a brand of radio that supports a variety of communication protocols, including ZigBee, 802.15.4, and WiFi, among others.

The ZigBee protocol is a standard the same way that Bluetooth is a standard. Any manufacturer's device that fully supports the ZigBee standard can communicate with any other company's ZigBee device. So just as your Motorola Bluetooth headset can communicate with your Apple iPhone, a CentralLite ZigBee light switch can communicate with a Black & Decker door lock. How does this work? Well, just like a great cake, robust network protocols are all about *layers*.

Most modern network protocols employ a concept of layers to separate different components and functions into independent modules that can be assembled in different ways. We're not going to bother with a lot of network theory here, just enough for you to complete the tasks at hand.

Every network has a physical layer where signals are actually transmitted. For example, your computer may be connected via an Ethernet cable to the Internet. On the other hand, it may be going wireless with a WiFi connection, using radio signals to traverse the real world. That's all happening in the physical layer, and doesn't change a thing about, for example, what's going on at the application layer, which is where your web browser lives. Firefox doesn't care a whit if you switch from Ethernet to WiFi. It is protected by the interfaces between layers that allow each software and hardware module to change how it does its job, but still talk to the other layers in exactly the same way.

Another way to conceptualize this is to consider your car. You can drive over concrete highways, asphalt driveways, metal bridges, and dirt parking lots without changing vehicles. Your tires provide an interface between the vehicle layer and the road layer. It would work just as well if you were driving a motorcycle or an ice cream truck. Either layer can be changed out independently without affecting the other.

The network layer below ZigBee that supports its advanced features is known as IEEE 802.15.4. This is a set of standards that define power management, addressing, error correction, message formats, and other point-to-point specifics necessary for proper communication to take place from one radio to another. XBee-brand radios can be purchased with or without ZigBee. For example, the XBee Series 1 hardware—which we don't work with in this book (but do mention in Chapter 1)—supports 802.15.4 directly in its native form. ZigBee is a set of layers built *on top of* 802.15.4. These layers add three important things:

Routing
> Routing tables define how one radio can pass messages *through* a series of other radios along the way to their final destination.

Ad hoc network creation
> This is an automated process that creates an entire network of radios on the fly, without any human intervention. Pretty cool.

Self-healing mesh
> Self-healing is a related process that automatically figures out if one or more radios is missing from the network and reconfigures the network to repair any broken routes.

A ZigBee network is a little like a basketball team. Both are composed of various players, and each player specializes in certain types of actions. Without the different players, neither can function properly. Of course, ZigBee is not quite basketball. For one thing, the radios are not particularly tall. Also, there are really only three kinds of players, or device types. Every ZigBee network will have a single *coordinator device*. You can't call anything a network until you have at least two things connected. So every ZigBee network will also have at least one other player, either a *router device* or an *end device*. Many networks will have both, and most will be much larger than just two or three radios:

Coordinator
> ZigBee networks always have a single coordinator device. This radio is responsible for forming the network, handing out addresses, and managing the other functions that define the network, secure it, and keep it healthy. Remember that each network must be formed by a coordinator *and* that you'll never have more than one coordinator in your network.

Router
> A router is a full-featured ZigBee node. It can join existing networks, send information, receive information, and route information. Routing means acting as a

messenger for communications between other devices that are too far apart to convey information on their own. Routers are typically plugged into an electrical outlet because they must be turned on all the time. A network may have multiple router radios.

End device

There are many situations where the hardware and full-time power of a router are excessive for what a particular radio node needs to do. End devices are essentially stripped-down versions of a router. They can join networks and send and receive information, but that's about it. They don't act as messengers between any other devices, so they can use less expensive hardware and can power themselves down intermittently, saving energy by going temporarily into a nonresponsive *sleep* mode. End devices always need a router or the coordinator to be their parent device. The parent helps end devices join the network, and stores messages for them when they are asleep. ZigBee networks may have any number of end devices. In fact, a network can be composed of one coordinator, multiple end devices, and no routers at all.

Network Topology

In basketball, once the players are selected, they still need to assemble as a team. ZigBee networks are the same way. They can connect together in several different layouts or topologies to give the network its structure. These topologies indicate how the radios are logically connected to each other. Their physical arrangement, of course, may be different. There are three major ZigBee topologies, illustrated in Figure 2-3:

Pair

The simplest network is one with just two radios, or *nodes*. One node must be a coordinator so that the network can be formed. The other can be configured as a router or an end device.

 In general, projects that will never need more than a single pair of radios won't get much advantage out of ZigBee and should consider using the simpler Series 1 802.15.4 protocol radios discussed in Chapter 1.

Star

This network arrangement is also fairly simple. A coordinator radio sits at the center of the star topology and connects to a circle of end devices. Every message in the system must pass through the coordinator radio, which routes them as needed between devices. The end devices do not communicate with each other directly.

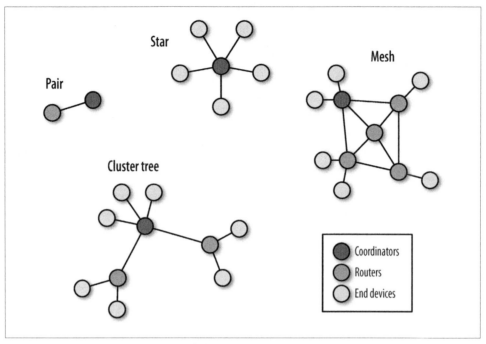

Figure 2-3. ZigBee pair, star, mesh, and cluster tree topologies

Mesh

> The mesh configuration employs router nodes in addition to the coordinator radio. These radios can pass messages along to other routers and end devices as needed. The coordinator (really just a special form of router) acts to manage the network. It can also route messages. Various end devices may be attached any router or to the coordinator. These can generate and receive information, but will need their parent's help to communicate with the other nodes.

Cluster tree

> This is a network layout where routers form a backbone of sorts, with end devices clustered around each router. It's not very different from a mesh configuration.

Addressing Basics

Almost every person has an address where he can be reached, usually one that is unique to him. Many people have more than one. We have mailing addresses, email addresses, phone numbers, passport numbers, and the list goes on. Each kind of address or identifier serves a slightly different purpose. It's the same with radios (see Table 2-1). If you want to send a ZigBee message, you need to know the address of the destination radio. Just like with people, each radio is known by several different addresses, each of which serves a purpose. For starters, each radio has a unique and permanently assigned 64-bit serial number. No other ZigBee radio on earth will have that same serial number.

Then there's a shorter 16-bit address that is dynamically assigned to each radio by the coordinator when it sets up a network. This address is unique only within a given network, but since it's shorter, many more of them can be manipulated in the very limited memory available on a ZigBee chip. Finally, each XBee radio can be assigned a short string of text called the node identifier. This allows the radio to be addressed with a more human-friendly name. Four out of five humans prefer a friendly machine.

Table 2-1. Address types

Type	Example	Unique
64-bit	0013A200403E0750	Yes, always and everywhere
16-bit	23F7	Yes, but only within a network
Node identifier	FRED'S RADIO	Uniqueness not guaranteed

PAN Addresses

In the United States, nearly every town has a Main Street. Thousands of different families live at, for example, 123 Main Street. We can tell them apart because while their street address is the same, their town or city is different. Each ZigBee network creates a virtual "city" in the same way, and labels that city not with a name but with a number, the Personal Area Network (PAN) address. This is another 16-bit address. There are 65,536 different PAN addresses available, each having the capability to generate another 65,536 16-bit radio addresses below it. In theory, therefore, this addressing scheme has room for more than 4 billion total radios, more than you'll ever need, no matter how ambitious a project you may have planned!

Channels

Even if all the addressing is perfect, your message still won't get through unless both radios are tuned to the same frequency. When the ZigBee coordinator picks a network PAN address, it also checks over all the available channels, typically 12 different ones, and picks a single one for that network's conversations. All the radios in that network must use the same channel. By default, XBee radios handle channel selection automatically so you usually don't need to worry about this, unless of course something goes wrong.

All Together Now

So for a message to get through from one radio to another, the radios need to be on the same channel and have the same PAN information, and the sending radio must know at least one of the receiving radio's addresses (see Figure 2-4). In addition, some networks have security protocols that require an exchange of keys; however, encryption and security protocols (discussed in Chapter 8) are not required for any of the projects we describe. For now, remember that you'll need a PAN and a radio address to get your messages through. You'll learn how these are set up in the next sections.

 You may run into terminology regarding ZigBee application layer addressing, including discussions of ZigBee *profiles*, *clusters*, and *endpoints*. You won't need these terms to do the projects in this book, so they are mentioned here only to reassure you that you can safely ignore them for now. We will cover application layer concepts in Chapter 8.

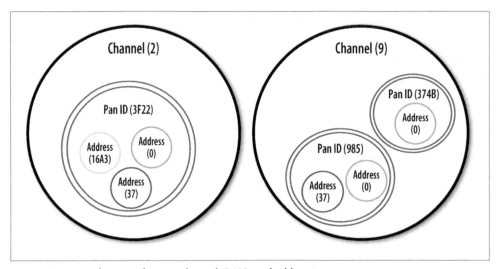

Figure 2-4. Venn diagram showing channel, PAN, and addressing

Hexadecimals

There's no question about it: if you want to use an XBee, you'll need to understand hexadecimal notation. Every time you set an XBee's address, configure one of its timers, or read the signal strength, the numbers you use are all formatted in base 16.

Relax! It's pretty easy and we're going to show you everything you need to know about these special numbers. If you've worked with computers at all, you've almost certainly seen these numbers, called hexadecimals, hex, or base 16 (these all mean the same thing).

Normally, we express numbers in base 10, counting with numerals from 0 to 9, then *carrying* to the next place to get 10 like this:

1...2...3...4...5...6...7...8...9...**10**

In decimal, there's a ones place, a tens place, a hundreds place, and so on. Let's take the decimal number 7,453:

7	4	5	3
thousands	hundreds	tens	ones
10^3	10^2	10^1	10^0

This probably looks pretty familiar. However, if you came from another planet and didn't know how to read decimals, you could multiply each number by its place to get the total value. $(7 * 1{,}000) + (4 * 100) + (5 * 10) + (3 * 1) = 7{,}453$. Hold that thought; this method will come in handy below.

You might be interested to learn that decimal is only one of many ways to write down numbers. Imagine how compact your notation would be if you could count from 0 to 15 before you had to carry to the next place, for example:

1...2...3...4...5...6...7...8...9... UH-OH!

We don't have a single numeral to express 10! We could make up a new squiggle and that would work fine, except it would be easier if we could use something already on the computer keyboard. So let's just use the letter A and say that stands for 10. We can then use the letter B to stand for 11, and so forth:

1...2...3...4...5...6...7...8...9...A...B...C...D...E...F...**10**

It may look weird but we didn't make a mistake at the end. F stands for 15, and then to express 16 we carried so we had 1 in the sixteens place and 0 in the ones place:

1	0
sixteens	ones
16^1	16^0

That's right, in hexadecimal the notation 10 means 16. To avoid confusion, we usually mark hexadecimal numbers in a special way, with a leading zero and a letter x, like this: 0x10. The leading zero and x don't mean that we're multiplying by zero or anything. They're simply a prefix, like the dollar sign, to let us know what kind of notation will follow. So if you see the hexadecimal 0x7E2 here's how to break it down:

7	E	2
two-hundred-fifty-sixes	sixteens	ones
16^2	16^1	16^0

So what number is this anyway? Remember that multiplication exercise we did with decimals? Let's try it with this unfamiliar hexadecimal. $(7 * 256) + (E * 16) + (2 * 1) = ??$ Oh bother, we need to translate that letter E into its decimal form. Let's do that right now: $(7 * 256) + (14 * 16) + (2 * 1) = 2{,}018$.

Try translating these decimals into hex. The first few are filled in to get you started:

10 = 0xA

16 = 0x10

17 = 0x11

18 = ___

160 = ___ (Hint: think of how many sixteens are in 160)

256 = ___

Now try translating these hexadecimals into decimals:

0xFF = ___

0x3 = ___

0x4B = ___

0x4C = ___

0x186A0 = ___

That last one is hard, so it's only fair to tell you now that both Windows and Macintosh have hex calculators. On the regular Windows calculator, change the View menu from Standard to Scientific. On the Macintosh calculator (Figure 2-5), change the View menu to Programmer. Now, click the Dec and Hex buttons to switch from decimal to hexadecimal notation. You were promised easy, and what could be simpler than clicking a button? Enjoy.

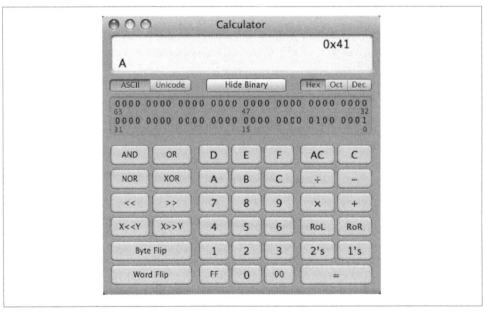

Figure 2-5. Mac calculator in programmer mode

XBee Firmware Updates

Your brain is brimming with facts, and your shiny new hardware sparkles with possibilities. The time has arrived to put your hands to work. Their first job will be to ensure that the right types and versions of the firmware are installed on your XBees. You'll be using the X-CTU program for this, so fire up the Windows operating system (or Linux; see "X-CTU in Linux" on page 33), then download and install X-CTU.

The X-CTU program and installation instructions are available at *http://www.digi.com/support/kbase/kbaseresultdetl.jsp?kb=125*. During the installation process, if you are asked if you want to download new firmware versions, go ahead and get them. In addition to X-CTU, you'll need to install the appropriate drivers for your XBee adapter board. Most adapter boards, such as the SparkFun XBee Explorer and the New Micros XBee Dongle, use FTDI drivers. The drivers and installation instructions are located at *http://www.ftdichip.com/FTDrivers.htm*. Windows may be able to discover the drivers on its own if you have Windows Update enabled and are connected to a network.

X-CTU in Linux

To use X-CTU under Linux, you'll need to first install Wine (*http://www.winehq.org*), which lets you run Windows applications under the X Window System. On a Linux system, you can usually install Wine using your Linux package manager.

Next, using Wine, download the X-CTU installer and run this command:

```
wine 40002637_c.exe
```

(If the filename is not *40002637_c.exe*, replace it with the name of the file you downloaded.)

Now create a symbolic link between the serial port that corresponds to your XBee and a Windows serial port, such as COM10:

```
ln -s /dev/ttyUSB0 ~/.wine/dosdevices/com10
```

The actual device filename (*ttyUSB0* in the example) will vary, so look at the output of the dmesg command shortly after you plug in the XBee adapter to see which device was added.

Next, launch X-CTU using a command something like:

```
wine .wine/drive_c/Program\ Files/Digi/XCTU/X-CTU.exe
```

Click the User Com Ports tab and type in the name of the Com port you created (such as COM10), then click Add. Unfortunately, you will need to do this each time you launch X-CTU, as it does not remember the custom Com ports.

Plug one of your XBee radios into your XBee adapter and connect the adapter to one of your Windows computer's USB ports. Launch the X-CTU application. You should see your XBee's USB connection listed under Select Com Port. Click on the appropriate port to select it, as shown in Figure 2-6.

The default settings in X-CTU will usually work for brand-new XBee radios that were configured at the factory. The easiest way to confirm that everything is set up correctly is to click on the Test/Query button once you've selected a COM port. If all goes well, you'll see a message that communication with the modem is OK and that gives you the modem type and firmware version, as shown in Figure 2-7.

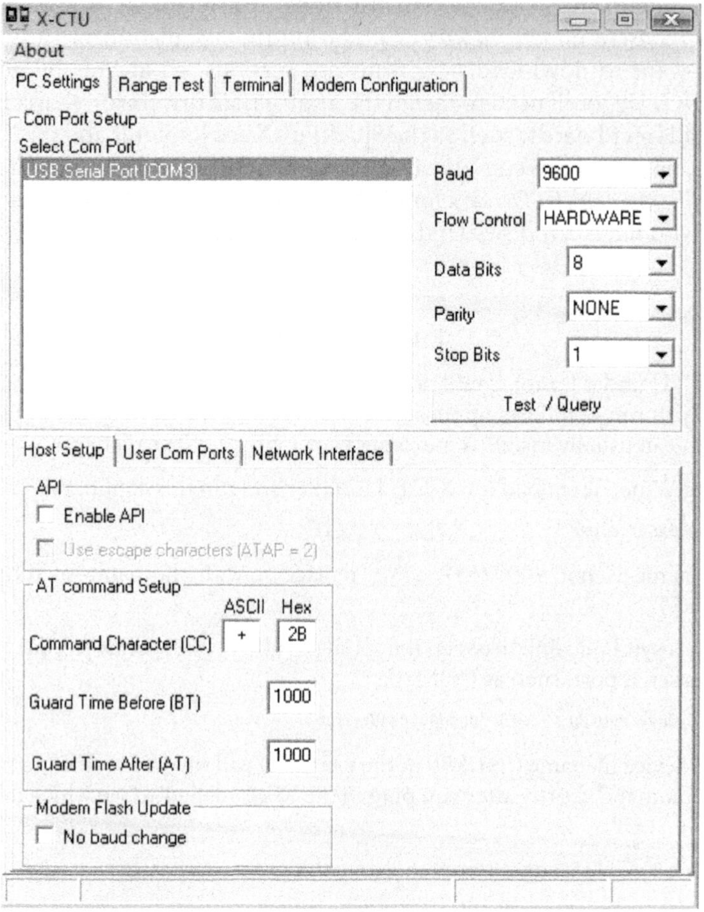

Figure 2-6. X-CTU starting screen

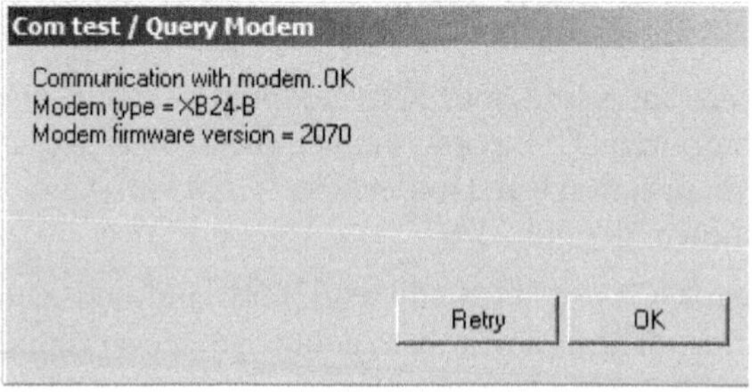

Figure 2-7. X-CTU test confirmation

If you get the message "Unable to communicate with modem," make sure your XBee is seated properly in its adapter, that it isn't too far forward or back by a pin, and that it wasn't inserted backward (see Figures 2-8 and 2-9). Also, check to make sure you selected the correct COM port. (If you suspect that your XBee may be using the API firmware, try checking the Enable API Mode box. API mode is covered in Chapter 5.) It's also possible that your XBee has been configured to a baud rate different from the default of 9600 baud. Try switching to one of the other baud rates and trying again. If you still can't get an OK response to the test, don't despair. Most of the time, your hardware is just fine. There are plenty of other fixes you can attempt. Check the Appendix for additional troubleshooting steps, or contact Digi for technical support at *http://www.digi.com/support*. Sometimes, just moving on to the next step helps with connection issues, so let's do that now.

Figure 2-8. An XBee misaligned and seated incorrectly. Note that one of the metal pins is showing ahead of the socket. This radio will not work until it is properly reseated.

Reading Current Firmware and Configuration

Now that you've tested the XBee for basic communication, you'll want to take a look at what firmware it's sporting and how that firmware is currently configured. Switch to the Modem Configuration tab, then click on the Read button under Modem Parameters. If all goes well, this will populate the window below with all kinds of useful information, as shown in Figure 2-10.

Figure 2-9. The XBee aligned and seated correctly in its adapter. All the metal pins are inserted into the sockets.

Note the "Download new versions" button. Use this button occasionally to have X-CTU check the Digi website for new versions of firmware (click "Download new versions," then click Web).

Linux Troubleshooting

If you're running X-CTU under Wine on Linux, you may see a dialog box that tells you the modem configuration file could not be found. This dialog will offer to download the latest configuration files from the website. If it fails:

1. Visit the Digi FTP site (*http://ftp1.digi.com/support/firmware/update*).

2. Next, look in both the *xbee_s2* (series 2) and *xbee_zb* (ZigBee firmware) subdirectories to find the firmware file that matches what you saw in Figure 2-7 (for example, *XB24-ZB_2041.zip* for a regular XBee, *XBP24-ZB_2041.zip* for an XBee-PRO).

3. Then, find the highest numbered *XB* (for XBee) or *XBP* (for XBee-PRO) firmware in the *xbee_zb* directory, sort by date, and download all of the most recent available ZIP files. For example, at the time of this writing, the most recent firmware files for the regular XBee were *XB24-ZB_2070.zip*, *XB24-ZB_2170.zip*, *XB24-ZB_2270.zip*, *XB24-ZB_2370.zip*, *XB24-ZB_2870.zip*, and *XB24-ZB_2970.zip*.

4. In X-CTU, go to the Modem Configuration tab, click Download New Versions, and use the File option to install each of the files (one at a time, unfortunately) you just downloaded.

5. Click the Read button again, and X-CTU should recognize your XBee.

The Modem type listed needs to be XB24-ZB or XBP24-ZB. Modem types XB24-B, XBP24-B, XBP24-SE, and XB24P-SE can be updated to the ZB firmware. If you see another model listed when you Read from the radio, it may not be the correct hardware, in which case it will not work with this book. Chapter 1 has information on where to obtain the correct hardware.

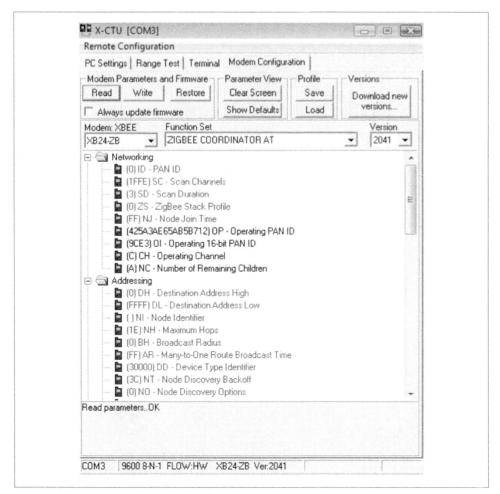

Figure 2-10. XBee coordinator AT configuration

If you get a dialog box that says there's a problem (Figure 2-11), with a suggestion for pressing the XBee reset switch, try gently pulling the XBee out of its socket on the adapter and reseating it. Be sure to wait 10 seconds for X-CTU to recognize the radio; after it does, it will close the dialog box on its own.

Doing this while the adapter is still plugged in effectively resets it. The Digi Evaluation Board has a reset button, so if that's your adapter, simply press the button.

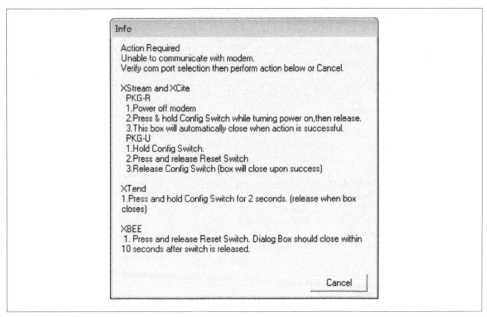

Figure 2-11. If you get this message, you can reset the XBee by gently pulling it out of its socket and reseating it

Let's configure the first XBee:

1. The class of radio modem is shown under Modem: XBEE (Figure 2-10). For everything we do in the main part of this book, it should be XB24-ZB (or XBP24-ZB if you're using the higher-power PRO version of the radios). If it's set to something else, select either of these options from the menu.

2. Under Function Set, you'll see a list of different firmware that can be loaded for this class of radio modem. To start, we'll be configuring one coordinator radio and one router, both in AT command mode. So for your first radio, if it's not already selected, choose ZIGBEE COORDINATOR AT for your function set. Any version 2070 or greater should be fine; in general you want the highest-number version (they're hexadecimals) listed for that particular function set.

3. Click on the Write button to program your radio with the coordinator firmware. For later reference, use a piece of tape or a small sticker to identify this radio as the coordinator.

 If you get a Windows error such as "Could not open output file. System error. Access denied.", check that your account has administrator access.

Once you've installed your first radio with the coordinator AT command software, gently remove that radio from the adapter and carefully seat a second radio in the same adapter. Click on Read in the Modem Configuration screen to see what firmware is on that radio, then select XB24-ZB (or XBP24-ZB for PRO radios), ZIGBEE ROUTER AT, and the highest-number version available. Any version 2270 or greater should be fine. Again, click on Write to program your second radio with the router firmware. Mark the router radio as well to identify it.

 If your radio has API firmware and you had to check the Enable API box on the PC Settings tab, when you switch to AT command firmware, the last step of the update may fail with a message about an "Error Setting AT Parameters" (Figure 2-12). You can safely ignore this error, change back to the PC Settings tab, uncheck the Enable API box, and then select the Modem Configuration tab and Read in the Modem Parameters again. Phew!

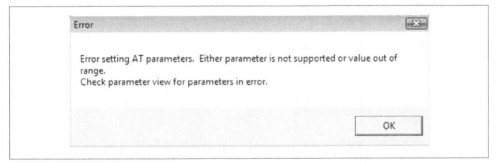

Figure 2-12. Error when switching from API to AT firmware

In addition to selecting firmware, you can use X-CTU to configure your radios' settings. Because you may not have full-time access to Windows and X-CTU, it's a good idea to learn how to change these settings with a regular serial terminal program. We'll start by setting up your XBees in this way.

Configuring XBee

Any time you aren't able to use X-CTU, you can configure any XBee that's in AT command mode by using a serial terminal program. In the previous chapter we covered a variety of serial terminal software. Here we'll show two different programs, Tera Term for Windows and CoolTerm for Macintosh (CoolTerm also works on Windows).

A Serial Terminal Program for Linux

If you're on Linux, you'll probably find picocom (*http://code.google.com/p/picocom/*) to be a suitable terminal program. The newer versions of picocom support local echo, which lets you see what you're typing. To use picocom, you'll need to compile and install it, then run it at the command line. For example, to connect to the first USB-serial port (if you've only got one XBee plugged into your Linux system, it will probably be this port), use:

```
picocom  --echo --imap crcrlf /dev/ttyUSB0
```

You can exit picocom by typing Ctrl-A followed by Ctrl-X.

Settings

No matter what program you use, you'll need to configure your software to use the communication settings shown in Table 2-2.

Table 2-2. Default XBee settings for serial terminal software

Baud	9600
Data	8 bit
Parity	None
Stop bits	1
Flow control	None
Line feed	CR+LF or Auto Line Feed
Local echo	On

Ports

You always need to select the USB port your XBee adapter is attached to. On Windows, this will probably be listed as one of the COM ports; on Macintosh, as a port with the word *usbserial* in the title; and on Linux, as a port with *ttyUSB* in the title. Many people figure out which port is right via trial and error. Honestly, this isn't a bad way to do it. The other option is to remove the XBee adapter from the USB port and see which port name disappears from your port list. The port name that disappears is your XBee adapter. Windows users can also find a list of the active COM ports by selecting the Device Manager from the Windows Control Panel on the Start menu. Macintosh users

---

can see a list of ports by opening the Terminal program, typing `ls /dev/tty.*` (Linux users should use `ls /dev/ttyUSB*`), and then pressing the Return key.

Connecting from Windows

To begin using your XBee via Tera Term on Windows, plug the XBee adapter into one of your USB ports and launch the Tera Term application. Tera Term can generally be selected right from the Windows Start menu. The opening screen (Figure 2-13) will prompt for a new connection.

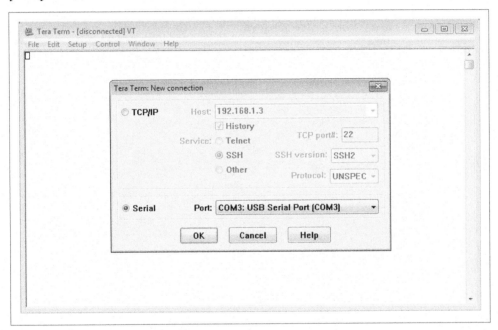

Figure 2-13. Tera Term opening screen

Select Serial on the "New connection" screen, then choose the port that is connected to your XBee adapter. Click on OK and you should see a blank Tera Term window. Choose Terminal from the Setup menu. In the dialog box that's presented (Figure 2-14), choose CR+LF for New-line Receive and check the "Local echo" box. Click OK to close this panel.

Next, select Serial from the Setup menu to confirm that the communication settings are correct. You've already selected your port, and the default of 9600 baud, 8 data bits, no parity, one stop bit, and no flow control will be perfect (Figure 2-15). Click OK to close the panel.

If you want to permanently save this setup, choose Save Setup… from the Setup menu and click the Save button. You're ready to get started with configuring your XBee!

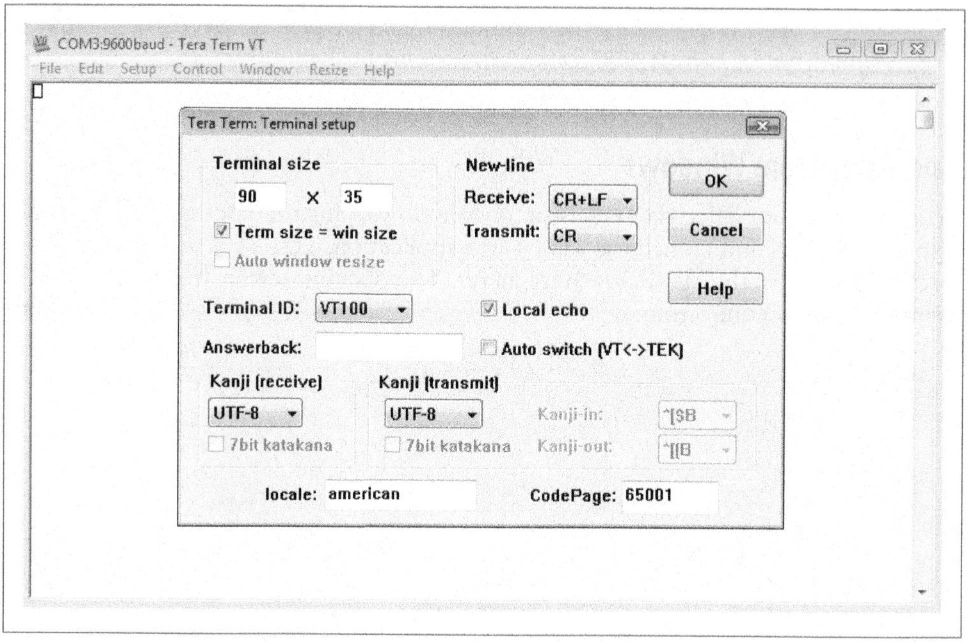

Figure 2-14. Tera Term terminal setup

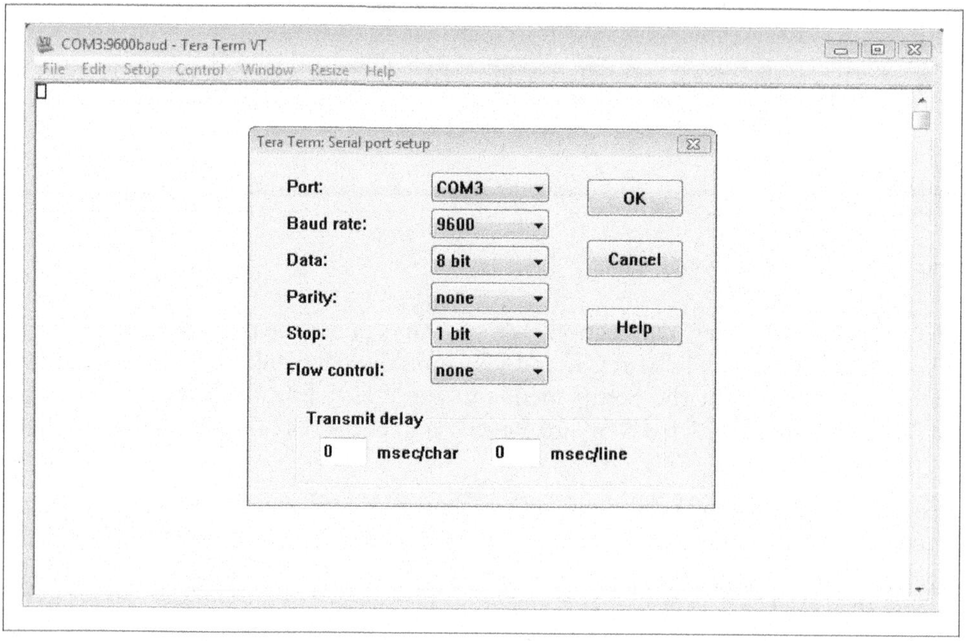

Figure 2-15. Tera Term serial port setup

To confirm that your XBee is connected properly, you can try putting it into command mode. Type three plus signs in a row, but *don't* press Return, just wait a moment after entering them:

+++

If you don't get a response, try typing the three plus signs again. Make sure you don't type anything before or after them. You should see an OK response (Figure 2-16). If you do, congratulations—you're successfully connected to your XBee! Skip to "Command Mode and Transparent Mode" on page 46 to continue.

 Your radio *requires* one second of guard time before and one second of guard time after you type the +++ or it won't go into command mode. Begin by typing nothing for at least one second, then type the three plus signs, and then don't type anything else for at least one second more. Remember, don't press the return key! That counts as typing something and will prevent you from going into command mode.

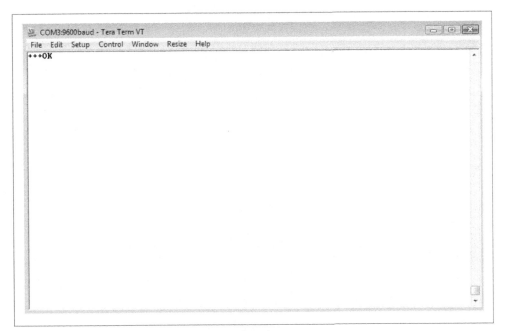

Figure 2-16. Tera Term with XBee in command mode

Connecting from Macintosh

To communicate with the XBee from the Macintosh OS, we will use an open source program called CoolTerm (Figure 2-17). CoolTerm is in ongoing development, so it will probably evolve rapidly and may behave somewhat differently or offer new features by the time you read this. Once you've downloaded and installed CoolTerm (*http://freeware.the-meiers.org*), double-click on the application to launch it. Click on the Options button to display the current settings (Figure 2-18). Choose the port your XBee adapter is connected to from the Device list at the top of the screen. The port will probably have `usbserial` as part of its title. The defaults of 9600 baud, 8-N-1 packets, and None for flow control will be perfect for initial connections to XBee radios. You'll want to check the Local Echo box so you can see the commands you're typing on-screen. Click the OK button to save your settings.

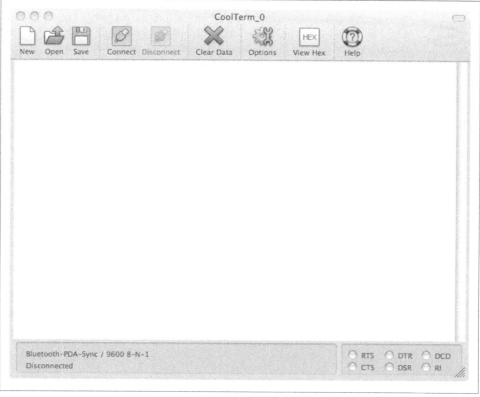

Figure 2-17. CoolTerm window

To open the serial connection, press the Connect button. Now you can put your XBee into command mode by typing three plus signs (**+++**) into the lower window. Don't press Return! The XBee should respond with OK about a second later. You'll see this response in the CoolTerm window (Figure 2-19).

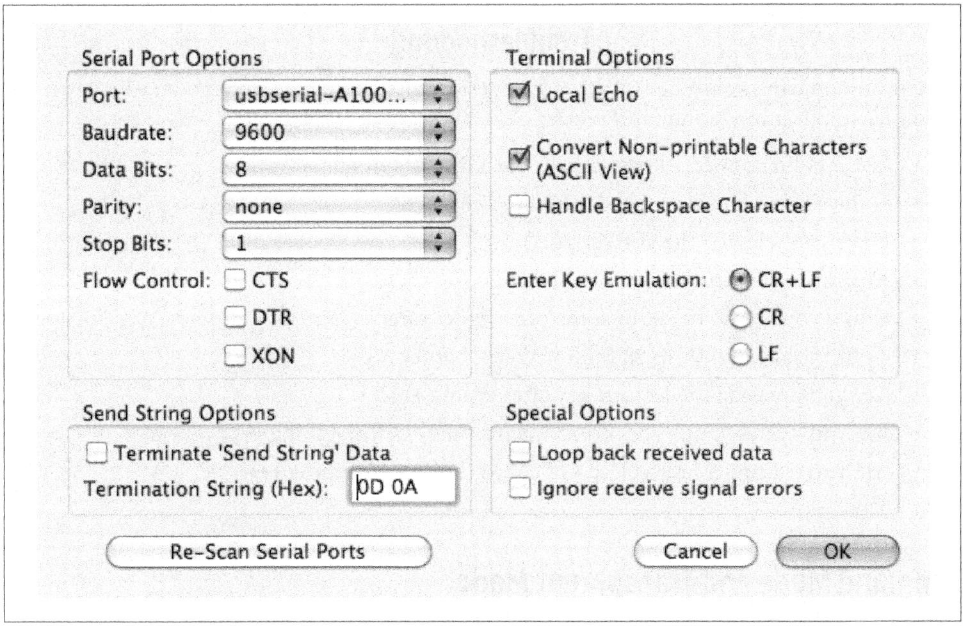

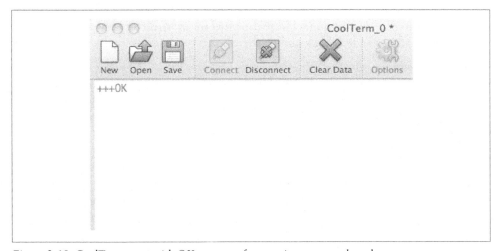

Figure 2-18. CoolTerm settings; select your port and check the Local Echo box

Figure 2-19. CoolTerm +++ with OK response for entering command mode

If you don't see the OK, check to make sure you've selected the correct port on the Options screen, and that you connected using the Connect button. Also make sure that you entered only +++. Don't press Return! The radio must get only the three plus signs or it won't go into command mode.

Troubleshooting

Here are some things to check if you aren't getting an OK response from your radio when you try to put it into command mode:

- Is the radio connected properly to the XBee adapter?
- Is the XBee adapter plugged into the computer?
- Have you selected the correct port?
- Are you communicating at 9600 baud?
- Could your radio be set to some other baud rate?
- Could your radio be in API mode? See Chapter 5.
- Are you pressing the Return key after typing +++?
- Are you waiting a full second of "guard time" before typing +++?
- Are you waiting a full second of "guard time" after typing +++?

Command Mode and Transparent Mode

All XBees communicate over radio with each other in the same way. However, they can use their local serial connection in two very different ways. Radios configured for API mode utilize a data-enveloping format that's great for computers talking to each other but is not easily human-readable. We'll be covering this in a later chapter. XBees that are configured to use "AT" commands are designed for more direct human interaction. AT-configured radios switch back and forth between two modes:

Transparent mode
> This is the default state for XBee radios using AT firmware. It's called transparent because the radio simply passes information along exactly as it receives it. Transparent mode is used to send data *through* the XBee to a remote destination radio. When data is received, it is sent out through the serial port exactly as it was received. What you send is what they get. Very simple.

Command mode
> Sometimes we don't want to send any data at all. Instead, we want to talk directly to the local radio, perhaps to ask about its configuration or alter the way it behaves. In this case we want to talk *to* the radio rather than through it. Rather than passing along what we type, the radio should stop, listen, and react. This is called command mode.

Table 2-3 summarizes these modes.

Table 2-3. Transparent versus command mode for AT radios

Transparent mode	Command mode
Talk *through* the XBee	Talk *to* the XBee itself

Transparent mode	Command mode
Any data can be sent through	Only responds to AT commands
Default state	+++ to enter mode
Wait 10 seconds to return to this mode	Times out after 10 seconds of no input

AT-configured XBees are normally in transparent mode. To get a radio to switch into command mode, we must issue a unique string of text in a special way. This is where those three plus signs come in (Table 2-4). When the radio sees a full second of silence in the data stream, followed by +++ and another full second of silence, it knows to stop sending data through and start accepting commands locally. (It's very unlikely that this particular combination would appear in the serial data by chance.) Once the radio is in command mode, it listens for user input for a while. If 10 seconds go by without any user input, the XBee automatically drops out of command mode and goes back into transparent mode.

Table 2-4. Entering command mode

Guard time silence	Command sequence	Guard time silence
One second before	+++	One second after

Remember that you must *not* press Return or Enter after typing the +++ because it will interrupt the guard time silence and prevent you from going into command mode!

AT Commands (Are Your Friend)

The AT commands that XBee radios use for interactive setup are a descendant of the Hayes command set that was originally developed for configuring telephone modems. The Hayes command set was never a codified standard, but many other modem manufacturers styled their command set after Hayes and today a variety of communications devices use the same format to accept configuration messages from serial connections.

You *always* need to press Enter or Return after issuing an AT command. Now just to be clear, the deal is to *never* press Enter after +++ and *always* press Enter after your AT command. You'll probably make mistakes with this at first, but it will come naturally soon enough.

Every AT command starts with the letters "AT," which stands for "attention." The idea is to get the attention of the device, in this case our XBee radio. The AT is followed by two characters that indicate which command is being issued, then by some optional configuration values. Here's an example:

```
ATID 1966<CR>
```

Don't type the <CR> literally. You just need to add a carriage return at the end, usually by pressing the Return key on the keyboard. It's a pretty simple structure that will be clear once you've issued a few commands. Here are some basic ones:

AT

> When the AT command is issued by itself, the radio will simply return OK. This is like asking "Are you there?" and the radio replying "Yup!" If you type **AT** and press Return, and don't see an OK in response, you've probably timed out of command mode and will need to type the **+++** to go back into it. This will happen a lot at first, but eventually you'll get used to the timing.

ATID

> Typing **ATID** by itself will show you the Personal Area Network ID that is currently assigned to the radio. PAN addresses define the network that a radio will attach to, using a hexadecimal number in the range 0x0–0xFFFFFFFFFFFFFFFF. Adding an address after the ATID command will assign a new PAN address to the radio. This is demonstrated below.

ATSH/ATSL

> Each XBee radio has a 64-bit serial number that serves as a permanent address that's unique to it in the world. The serial number address is split into two parts, a high part and a low part. This is because a single register is not big enough to hold the whole address. It can't be changed, so while typing **ATSH** or **ATSL** will show you the high and low parts of that serial number respectively, adding any address information after this command will cause an ERROR response.

ATDH/ATDL

> These show or assign the destination address that the local radio will send information to. Typing **ATDH** will *show* you the current high part of the destination address, while putting address information after ATDH will *set* a new high part to the destination address.

ATCN

> This command will drop you out of command mode immediately, returning the radio to transparent mode. You can also type nothing for 10 seconds and the radio will drop out of command mode automatically.

ATWR

> This writes the complete current configuration to firmware, so that the next time the radio powers up it has the new configuration. **ATWR** is similar to a Save command on a computer that writes your document to the hard drive so it's stored even after the computer is turned off.

ATMY

> This command shows you the current 16-bit address. The coordinator assigns this address dynamically so it can be displayed (but not set) for the Series 2 ZigBee radios.

Using AT Commands

Issuing any of these AT commands is very easy on both Windows and Macintosh. Here are the steps:

1. Use **+++** to ask the radio to go into command mode.
2. Wait for an OK response.
3. To *read* a register, type an AT command such as **ATID**, followed by a carriage return.
4. To *set* a register, type an AT command followed by the parameter you want to set, like this: **ATID 1966**, followed by a carriage return. The space before the parameter is optional so this also works: **ATID1966.**

Figure 2-20 shows how this looks in Windows.

Figure 2-20. Reading, setting, and then rereading a register in Tera Term

In CoolTerm on a Macintosh, the procedure works pretty much the same. After putting the radio into command mode with **+++**, issue an AT command by typing it in the window, followed by a carriage return.

The results from querying with **ATID**, setting **ATID 1966**, and then querying with **ATID** again are shown in Figure 2-21.

Now that you know how to connect to your XBee, put it in command mode, and issue AT commands, you're ready to configure two radios to chat with each other. Let's start that scintillating conversation.

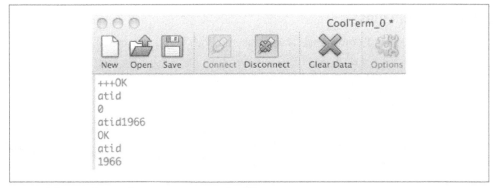

Figure 2-21. Reading and setting registers in CoolTerm

Basic ZigBee Chat

Networks are all about connections, so configuring a single radio doesn't qualify as making a network because it's not connected to anything. You need at least two radios to make a network, so here's what you'll need to create your first project—a simple ZigBee chat session.

Parts

- One XBee Series 2 radio, configured as a ZigBee Coordinator AT (Digi: XB24-Z7WIT-004, DK 602-1098-ND).
- One XBee Series 2 radio, configured as a ZigBee Router AT (same as previous).
- Two XBee USB adapter boards (SFE: WRL-08687).
- Two computers, each running a serial terminal program, or one computer running two different serial terminal programs. Using two computers is less confusing, so find a buddy if you can.

Addresses

Every XBee radio has a 64-bit serial number address printed on the back (Figure 2-22). The beginning or "high" part of the address will be 0013A200, Digi's pre-assigned range of address space. The last or "low" part of the address will be different for every radio. It will look something like this: 4052DAE3.

Write down your coordinator and router addresses so you can refer to them later:

Coordinator address	Router address
0013A200 _____	0013A200 _____

Figure 2-22. Back of XBee showing 64-bit address

Coordinator

Start with the XBee ZIGBEE COORDINATOR AT radio you configured earlier in this chapter. Remember that every ZigBee network must have one coordinator radio—and only one coordinator radio—so that the network can be properly defined and managed. Use your serial terminal program and AT commands (or X-CTU if you have access) to configure the coordinator radio with the settings in Table 2-5.

Table 2-5. Coordinator setup for paired chat

Function	Command	Parameter
PAN ID	ATID	**2001** (any address from 0 to FFFF will do)
Destination address high	ATDH	**0013A200**
Destination address low	ATDL	**<see your recorded Router Address>**

When you're finished, check your work by reissuing each AT command without any parameter so the radio can show you the addresses it's using (Figure 2-23).

As a final step, use the **ATWR** command to write the new configuration to your radio's firmware so it's saved for the next power-up.

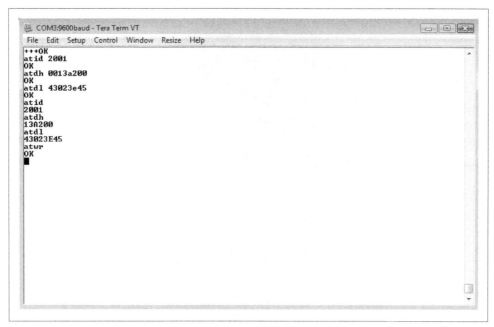

Figure 2-23. Setting and checking the coordinator radio

Here's what a session might look like:

```
+++
OK
ATID 2001
OK
ATDH 0013A200
OK
ATDL 43023E45
OK
ATID
2001
ATDH
13A200
ATDL
43023E45
ATWR
OK
```

You should get an OK response after issuing each command to set parameters, and another OK response when you write the changes to firmware. If you don't get an OK response, most likely you took more than 10 seconds to issue the command and you've dropped out of command mode. This can happen quite frequently when you're starting out, but you'll get better at it as you go along. The other common mistake is not issuing the **ATWR** command to save your changes, then losing your configuration when the radio is powered down.

The command mode timeout can be changed to a longer value with **ATCT**, but it's best to wait on doing this until you are more comfortable with the radios so you don't set the timeout to an impossibly short value by accident.

Router

Replace the coordinator radio with the XBee ZigBee Router AT radio you configured earlier in this chapter. Use your serial terminal program and AT commands (or X-CTU if you have access) to configure the router radio with the settings in Table 2-6.

Table 2-6. Router setup for paired chat

Function	Command	Parameter
PAN ID	ATID	*2001* (must be the same for all radios on your network)
Destination address high	ATDH	*0013A200*
Destination address low	ATDL	*<see your recorded Coordinator Address>*

When you've finished configuring the radio, check your work by reissuing each AT command without any parameter so the radio can show you the addresses it's using (see Figure 2-24).

As a final step, use the **ATWR** command to write the new configuration to your radio's firmware so it's saved for the next power-up. Disconnect the XBee from the computer for now.

Two Computers

Get ready to chat. Connect your coordinator XBee via an adapter to one computer's USB port. Launch a serial terminal application, or use the Terminal tab in X-CTU. (Make sure you select the current port and configure the terminal application for the right baud rate and other settings.) Your router radio should be connected in the same way to the second computer, which should be running its own serial terminal application.

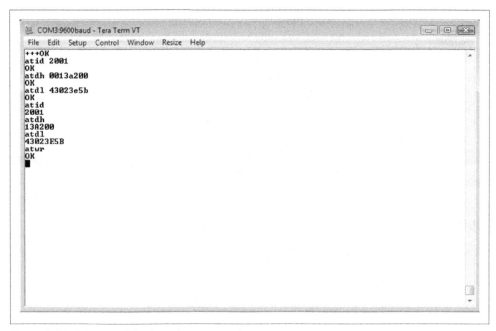

Figure 2-24. Setting and checking the router radio

One Computer

If you have only one computer, connect both radios to that computer's USB ports. Then choose two different terminal windows, like Tera Term and HyperTerminal on Windows, or CoolTerm and ZTerm on Macintosh. Pick one radio's port in one program and the other radio's port in the other program. Pretend that your first terminal program is one computer, and that your second terminal program is another one. Keeping all of this straight might make your head hurt a bit, but it's a valid test when you can't find a real second computer.

Chat

This is the moment you've been waiting for. If everything is set up properly, the text that you type in the serial terminal program on the first computer will be relayed to the second computer and appear on its serial terminal screen as well. Give it a try.

> Remember that chatting will happen only when the radios are in transparent mode. If you are in command mode, type **ATCN** and press Return, or simply wait 10 seconds for command mode to time out.

Troubleshooting

If everything works perfectly the first time around, that's GREAT! However, experience shows that it sometimes takes a few tries to get everything right. You've just set up a pretty complex system. Don't despair if your chat doesn't work right away. In almost every case, there's nothing wrong with any of your hardware or even with most of your setup. It takes only one wrong parameter to throw a wrench in the works. Learning how to find that wrench and fix it is an essential skill, so here are some tips on what to try if things don't work at first:

1. Start with the simple stuff. Make sure your radios are seated properly in the adapter boards and that all the USB cables are plugged in the way they should be.

2. Check that each radio is responding properly in the terminal window by trying to use **+++** to put it into command mode. If you don't get an OK back, check your port selection, baud rate, and the other settings until you find the reason the radio is not communicating properly. (If you accidentally configured either radio with API firmware, it will not respond and you will need to change firmware to the AT version in X-CTU.)

3. If both radios are responding, use AT commands to check the settings. The most common problems are: not using the same PAN ID on both radios, not setting the destination address on each radio to the serial number of the *other* radio, and not saving the settings properly.

4. If the settings all seem to be correct, check to make sure that you have the coordinator firmware on one radio and the router firmware on the other radio. You can use the **ATVR** command to show which firmware is in use. The coordinator's version number will begin **20xx** and the router's version number will begin **22xx**. If you see other values, go back to X-CTU and load the proper firmware.

5. Sometimes the radios will be perfectly configured and connected, but your router will have joined a different network. This often happens in classroom situations, where many people are using the same PAN ID in the same room. Try using different PAN IDs for each pair.

6. A router will normally find the coordinator on the same PAN and join itself to the network. Very occasionally this won't happen properly. You can force each radio to rebuild its networking setup from scratch by issuing a network reset with **ATNR**. Do this on both radios, then recheck to see if they are now connecting properly.

7. Sometimes setting both radios back to factory defaults and reconfiguring them will flush out a bad setting that was left over from a previous setup, or an unrecognized typo. The **ATRE** command will wipe out your radio's custom configuration and leave the firmware set cleanly to factory defaults. Follow it with the **ATWR** command to write those defaults to the firmware, then go back to the configuration steps and try putting in your settings again.

8. Don't forget that Digi's technical support is a great resource. While they needn't be your first step, if you're really stuck they can help you confirm that your radios are working properly at the hardware level. There are also a number of great online resources and forums you can read for ideas and where you can ask for more help. Check the resource guide in the Appendix.

Success!

When you *do* get the chat working, this is cause for a *major* celebration. Dance the hokey-pokey, sing Norway's national anthem, eat pudding, or do all three at the same time. Your very first ZigBee network is up and running!

CHAPTER 3

Build a Better Doorbell

Now that the wheels are in motion, you're probably itching to create something practical. Let's get on with it. We'll briefly introduce the Arduino microcontroller system, with basic instructions for configuration and use. Since this isn't an Arduino book, we'll cover only what you need to know to get this project done. If you're new to Arduino and interested in learning more, ample references to other books and sites will be provided to help you learn whatever is beyond our fairly narrow scope. After getting up to speed on basic serial concepts and simple protocols, we'll execute a series of doorbell projects that build in creative complexity as you gain skill. Knock, knock. Who's there? Wireless interaction!

ZigBee and Arduino

Arduino and XBees can work extremely well together in wireless sensor systems. They are both great prototyping tools.

About Arduino

Arduino (Figure 3-1) is an open source microcontroller system that's very popular with prototypers, do-it-yourself enthusiasts, interaction designers, and educators. The system is designed to be easy to learn, easy to use, flexible, and fast to develop with. Microcontrollers are little computers that do specific jobs, such as taking input from switches and sensors and then deciding whether to turn on a light or ring a bell. They're widely used in portable devices, including the types you might want to use in a wireless sensor network. Here's how the Arduino project describes itself on its website (*http://www.arduino.cc*):

> Arduino can sense the environment by receiving input from a variety of sensors and can affect its surroundings by controlling lights, motors, and other actuators. The microcontroller on the board is programmed using the Arduino programming language (based on Wiring) and the Arduino development environment (based on Processing). Arduino

projects can be stand-alone or they can communicate with software running on a computer (e.g., Flash, Processing, MaxMSP).

The boards can be built by hand or purchased preassembled; the software can be downloaded for free.

Figure 3-1. The basic Arduino USB board

In addition to Arduino, there are a slew of other microcontroller systems available, including the PIC chip, BASIC Stamp, Beagle Boards, and more. If you happen to prefer one of those platforms, simply use this book's examples and code as a guide.

We will see in upcoming chapters that the XBee is capable of doing some sensing and actuation without an external microcontroller, yet we're already starting with an additional piece of equipment. Actually, it's for a good reason. External microcontrollers bring several important advantages to a wireless project, including:

Local logic

While the basic XBee radios can be a source of sensor data or a trigger for local output, they can't be programmed to perform logical information processing. If your sensor or device needs local decision-making, you will almost certainly want to add a microcontroller to handle those processes.

Additional input/output lines

Series 2 XBee hardware comes with 10 digital input/output lines, 4 of which can be configured for analog (variable) input. While using an XBee, you can configure the basic Arduino to use up to 17 digital input/output lines, 6 of which can take analog input while 6 others offer hardware support for analog output. If you have extensive input or output needs, an external microcontroller may be just the thing.

Fast prototyping

It is generally much easier to deploy and test a solution using a simple, high-level development system like Arduino than to mess with the XBee's application programming interface and data envelope frames. Even if you just want to do simple input/output on the XBee module, adding an external microcontroller will probably save you time as you try out your initial idea. If everything pans out, you can always slim your project down later.

Lots of connection options

With the help of an Arduino, your XBee can drive large motors, interact with GPS modules, drive LCD display screens, store data in local memory banks, and interact directly with the Internet via WiFi or your mobile phone. Working together, the possibilities are limitless.

Arduino Basics

Here's how to get ready to work with the Arduino microcontroller system.

Buying an Arduino

Arduino hardware comes in many flavors. The basic Arduino as of this writing is the *Uno*. This model supports 13 digital input/outputs along with 6 analog inputs. It can run off of USB power or via an external "wall wart" power supply. The onboard microcontroller supports up to 32K of program code with 2K of RAM. This may not seem like a lot, but in 8-bit microcontroller terms it's probably more than most prototypers need. The main Arduino website hosts an exhaustive list of sources at *http://www.arduino.cc/en/Main/Buy*, or you can find it at Maker Shed, SparkFun, Adafruit, and many other online retailers.

There are plenty of other options if your project has special needs. For example, the Arduino *Mega* is good for very big jobs. It has 54 digital input/outputs and 16 analog inputs, along with 4 hardware serial connections. Should you want to go small, check out the Arduino *Mini*, which omits USB and female headers to allow a much smaller form factor, though at the expense of some prototyping ease. You'll find plenty of Arduino clones available too, all of which are configured a little differently to suit particular needs and tastes. If you feel bewildered by the options, the *Uno* is a fine choice for getting started. All the examples in this book are based on it.

Don't forget the cable

You'll want a USB cable for programming your Arduino board. For the *Uno* or *Mega*, you need the easily obtained A-to-B-style USB cable. Radio Shack carries these and you can also find them online at places like OutletPC.com, where they often are on sale for less than $1.

Downloading the software

The Arduino is programmed using an open source application that runs on your computer. This is known as the IDE (or integrated development environment) and you can download it for free directly from the Arduino website's software area (*http://www.arduino.cc/en/Main/Software*). There are versions available for Macintosh, Windows, and Linux. Download the appropriate version for your computer. You'll find a basic guide to getting started at *http://arduino.cc/en/Guide/HomePage*.

Using the Arduino IDE

The Arduino IDE (Figure 3-2) is split into three areas. The blue area at the top of the window features a toolbar of buttons that control program behavior. The white area in the middle is where you enter and modify code. The black section at the bottom of the window is where status messages appear, and where you should look for error messages that can help you debug your code.

As described in the online Arduino guide, the toolbar buttons perform the following functions:

▷	Verify/Compile	Checks your code for errors.
◻	Stop	Stops the serial monitor or removes the highlight from other buttons.
▯	New	Creates a new sketch (what Arduino programmers call their programs).
⇧	Open	Presents a menu of all the sketches in your sketchbook (the Arduino program directory). Clicking one will open it within the current window.
⭳	Save	Saves your sketch.
⇨	Upload to I/O Board	Compiles your code and uploads it to the Arduino board.
▤	Serial Monitor	Opens the serial monitor.

Selecting the board and port

To connect to your Arduino board, you must plug it into your computer using a USB A-to-B-style cable. Next, select the model of your Arduino board from the Board menu. Finally, select your serial port from the Serial menu. On Windows computers, the serial port will be one of the COM ports. On Macintosh, the serial port will have a name that includes `usbserial`, followed by some identifying letters and numbers. Once you've selected your board type and port, you're ready to do some programming!

Code structure

The Arduino language is based on C/C++ and as such it shares a specific set of structures that have been simplified for people new to programming. A simple program might look something like this:

```
// variable definitions always come first
```

```
int ledPin =  13;


// The setup() method runs once, when the sketch starts

void setup()   {
  // initialize the digital pin as an output:
  pinMode(ledPin, OUTPUT);
}


// the loop() method runs over and over again,
// as long as the Arduino has power

void loop()
{
  digitalWrite(ledPin, HIGH);   // set the LED on
  delay(1000);                  // wait for a second
  digitalWrite(ledPin, LOW);    // set the LED off
  delay(1000);                  // wait for a second
}
```

A basic program begins with statements that declare the names, types, and initial values for named containers that are used throughout the program, also known as *global variables*. Next comes a section that begins with void setup(). Everything between the curly braces for this section is code that runs only once, right after the Arduino is powered up or reset. Typically this section contains procedures that get the Arduino board ready to do its work, like initializing pins, setting up serial ports, and anything else that needs to happen only once, on startup. Finally, there's a section that starts with void loop(). The code contained in the loop section, between its curly braces, runs constantly. In the example above, this code will turn on an LED light, wait a moment, then turn it off and wait a moment. That's one blink of a blinking light. As soon as the first blink is complete, the loop code runs again, meaning the light will blink on and off indefinitely. Sometimes there will be other sections following the loop. These describe additional functions that are typically called from the main loop, but also could be called from setup or by each other. For full information on getting started with Arduino programming, take a look at the longer explanation at *http://arduino.cc/en/Tutorial/Foundations* and the many examples that can be downloaded from *http://arduino.cc/en/Tutorial/HomePage*. You'll find the complete language reference at *http://arduino.cc/en/Reference/HomePage*. These resources are also available from the Help menu in the Arduino IDE. Figure 3-3 shows the Arduino board in detail.

Buying electronics supplies

Arduino projects almost always require additional components, such as switches, lights, sensors, knobs, wiring, or motors. Here are some resources where you can find the electronics goodies you need:

Figure 3-2. Arduino IDE programming software

Maker Shed (http://www.makershed.com/)

> Has lots of projects and some good components, including a kit specifically designed for use with this book.

Adafruit (http://adafruit.com/)
> Has a great collection of electronics kits with a small but useful selection of electronic components that are very appropriate for beginners.

SparkFun (http://www.sparkfun.com)
> Contains a wealth of terrific prototyping components, each well-explained for the amateur electronics enthusiast.

DigiKey (http://www.digikey.com)
> A complete resource for electrical engineering, DigiKey stocks half a million different components and can deliver any of them overnight. Take a deep breath before shopping because almost every part is available in a hundred different variations.

Jameco (http://www.jameco.com)
> Another electrical engineering resource, Jameco tends to have a more limited selection. This, along with its full-color catalog, can be a boon for those new to selecting components.

Mouser (http://www.mouser.com)
> Similar to DigiKey, Mouser carries a huge selection of parts. If you can't find it from one, check the other.

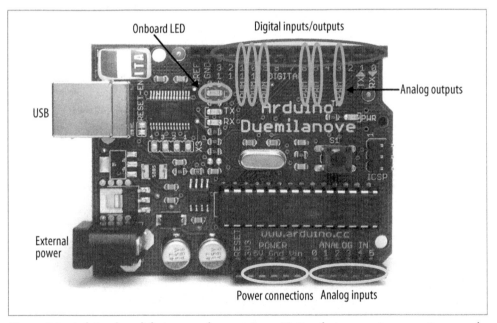

Figure 3-3. Arduino board features and connections. Notice that some pin connections can be configured for several different purposes.

Learning More About Arduino

There's lots to know about the Arduino system, more than we could possibly cover in this chapter. Luckily, there are plenty of resources available to ease your journey from novice to expert.

On the Arduino site:

- The Getting Started guide (*http://arduino.cc/en/Guide/HomePage*).
- The Language Reference area lists all the commands and shows how to use them (*http://arduino.cc/en/Reference/HomePage*).
- The Arduino Playground hosts a cornucopia of resources, including a wealth of completed projects and information on extending the basics with more advanced techniques (*http://www.arduino.cc/playground/*).
- The Hardware area lists most available boards as well as "shields" to extend them (*http://arduino.cc/en/Main/Hardware*).

On the Internet:

- The LadyAda Arduino Tutorial (*http://www.ladyada.net/learn/arduino/*).
- The NYU ITP Physical Computing tutorials for getting started (*http://itp.nyu.edu/physcomp/Tutorials/*) and the sensor wiki for understanding a variety of components (*http://itp.nyu.edu/physcomp/sensors/*).
- The SparkFun set of tutorials, from basic to iPhone (*http://www.sparkfun.com/commerce/tutorials.php*).
- The Sheepdog Guides Arduino Course (*http://sheepdogguides.com/arduino/FA1main.htm*).
- The Freeduino online index features links to guides and tutorials from all over (*http://www.freeduino.org/*).

In Arduino communities:

- The Arduino site forum is a good place to meet people, see if your question has already been answered, and if not, ask it (*http://www.arduino.cc/cgi-bin/yabb2/YaBB.pl*).
- Both SparkFun and Adafruit have very active forums as well (*http://forum.sparkfun.com/*; *http://forums.adafruit.com/*).

At hacker spaces:

- The local one in New York is NYC Resistor (*http://www.nycresistor.com/*).
- Hundreds more from Maui to West Bengal are listed on Hackerspaces (*http://hackerspaces.org/wiki/List_of_Hacker_Spaces*).

In books:

- *Getting Started with Arduino* (*http://oreilly.com/catalog/9780596155520/*) by Massimo Banzi (O'Reilly, 2008).

- *Practical Arduino: Cool Projects for Open Source Hardware* by Jonathan Oxner and Hugh Blemings (Apress, 2009).
- *Arduino Cookbook (http://oreilly.com/catalog/9780596802486/)* by Michael Margolis (O'Reilly, 2011).
- *Physical Computing: Sensing and Controlling the Physical World with Computers* by Dan O'Sullivan and Tom Igoe (Thomson, 2004) is a great reference to the process of building humanized interactive systems.

Finally, here's a brief history of the Arduino:

- Clive Thompson's great story about Arduino and open source hardware, "Build It. Share It. Profit. Can Open Source Hardware Work?" *http://www.wired.com/techbiz/startups/magazine/16-11/ff_openmanufacturing*).

Connecting to Arduino

The solderless breadboard and XBee breakout boards described in Chapter 1 provide an easy way to link your Arduino board to an XBee. While the XBee has many pins, it takes only four of them to create a working connection so that your Arduino can communicate wirelessly, using its built-in serial communications protocol.

Remember that the XBee pins are spaced 2 mm apart, so the XBee can't be placed directly into a breadboard. A basic breakout board is the least expensive adapter for connecting to an Arduino. You can also use an XBee Explorer as a breakout board, but keep in mind that the pins are arranged differently. The four connecting wires will provide power, electrical ground, transmit, and receive. Table 3-1 shows the pin connections between Arduino and XBee, and Figure 3-4 shows them on an XBee breakout board.

Table 3-1. Pin connections between Arduino and XBee

XBee	Arduino
VCC or 3.3 V	3V3
TX or DOUT	RX or 0
RX or DIN	TX or 1
GND	GND

Remember, if you are using the XBee Explorer you'll connect the same pins but their physical layout will be different, as shown in Figure 3-5.

After plugging your XBee into a small breadboard, you can use different colors of hookup wire to make the connections between your Arduino and XBee (see Figure 3-6). Once connected, the Arduino uses serial commands to send information out via the XBee, and to read in any information that's received. This is how our doorbells will operate.

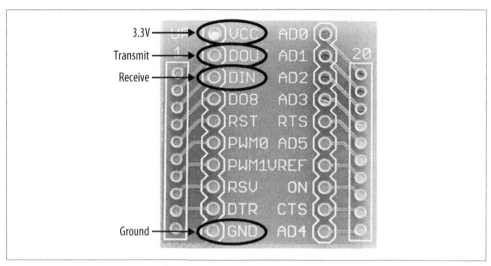

Figure 3-4. Breakout board pins for serial connection to Arduino

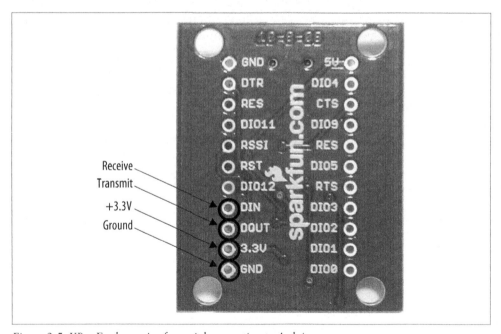

Figure 3-5. XBee Explorer pins for serial connection to Arduino

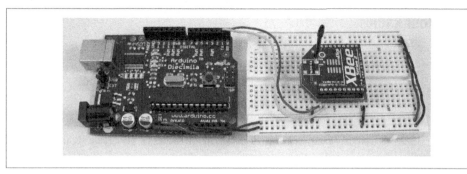

Figure 3-6. Arduino connected to an XBee radio, using a solderless breadboard and hookup wire

Doorbell Projects

You are now ready to create your first stand-alone wireless systems. The projects in this chapter use pairs of radios to help you learn networking basics. In a couple of chapters, you'll be creating much larger networks, using the skills you'll learn here.

Parts

- Two solderless breadboards (MS MKKN2, AF 64, DK 438-1045-ND, SFE PRT-09567)
- Hookup wire or jumper wire kit (MS MKSEEED3, AF 153, DK 923351-ND, SFE PRT-00124)
- Two Arduino boards (MS MKSP4, SFE DEV-09950, AF 50)
- USB A-to-B cable for Arduinos (AF 62, DK 88732-9002, SFE CAB-00512)
- An LED (try the 5 mm size, and make sure you don't buy any surface mount (SMT) parts) (DK 160-1707-ND, RS 276-041, SFE COM-09590)
- One 10K ohm resistor (DK P10KBACT-ND, SFE COM-08374)
- One momentary switch for input (DK EG2025-ND, RS 275-618, SFE COM-09179)
- One buzzer for output (DK 102-1621-ND, RS 273-060)
- One XBee radio (Series 2/ZB firmware) configured as a ZigBee Coordinator AT mode (Digi: XB24-Z7WIT-004, DK 602-1098-ND)
- One XBee radio (Series 2/ZB firmware) configured as a ZigBee Router AT mode (Digi: XB24-Z7WIT-004, DK 602-1098-ND)
- Two XBee breakout boards with male headers and 2 mm female headers installed (AF 126 [add SFE PRT-00116], SFE BOB-08276, PRT-08272, and PRT-00116)
- XBee USB Serial adapter (XBee Explorer, Digi Evaluation board, or similar) (AF 247, SFE WRL-08687)
- USB cable for XBee adapter (AF 260, SFE CAB-00598)
- Wire strippers (AF 147, DK PAL70057-ND, SFE TOL-08696)

Prepare Your Radios

Every ZigBee network needs one and only one node configured as a coordinator. The other nodes can be configured as routers (or end devices):

1. Follow the instructions under "Reading Current Firmware and Configuration" on page 35 in Chapter 2 to configure one of your radios as a ZigBee Coordinator AT.

2. Using the same instructions, configure your other radio as a ZigBee Router AT.

3. Label the coordinator radio with a "C" so you know which one it is later on. Label the router radio with an "R."

Connect Power from Arduino to Breadboard

1. Hook up a red wire from the 3.3 V output of the Arduino to one of the power rails on the breadboard (see Figure 3-7).

2. Hook up a black wire from either ground (GND) connection on the Arduino to a ground rail on the breadboard.

3. Hook up power and ground across the breadboard so that the rails on both sides are live.

 Make sure you are using 3.3 V power. The XBee will *not* run on 5 volts, and any more than 7 volts will permanently damage it.

XBee Breakout Board

Your XBee radio has pins spaced 2 mm apart. This helps keep the component small, but it means you can't use it directly on a standard 0.1″-spaced solderless breadboard. To mate it with the breadboard, you need to use a breakout board. Basic breakout boards have no other electrical components. Another option is to use certain XBee USB-serial adapters, such as the XBee Explorer, Adafruit XBee Adapter, or MCPros XBee Simple Board, all of which come with standard-spaced holes where you can solder on male headers. In this example, we'll just work with a basic breakout board:

1. Solder regular 0.1″-spaced male headers onto the two center rows of holes on your basic XBee breakout board. The male headers come in long strips, and must be cut down to fit the breakout board *before soldering*. It's a good idea to place the male headers into the breakout board and insert them into the breadboard, as this helps with stability while soldering.

2. Next, flip the board over and solder two strips of female 2 mm-spaced sockets onto the *opposite* side of the breakout board.

3. Test-fit the XBee into the female sockets, being careful not to bend its pins (see Figure 3-8).

Figure 3-7. Power connections on Arduino

Figure 3-8. Finished breakout board with XBee mounted

XBee Connections

1. With the XBee mounted on its breakout board, position the breakout board in the center of your breadboard so that the two rows of male header pins are inserted on opposite sides of the center trough.

2. Use red hookup wire to connect pin 1 (VCC) of the XBee to 3.3 volt power. See Figure 3-7.

3. Use black hookup wire to connect pin 10 (GND) of the XBee to ground.

4. Use yellow (or another color) hookup wire to connect pin 2 (TX/DOUT) of the XBee to digital pin 0 (RX) on your Arduino (see Figure 3-9).

5. Finally, use blue (or another color) hookup wire to connect pin 3 (RX/DIN) of your XBee to digital pin 1 (TX) on your Arduino. Figure 3-10 shows the connections to the XBee.

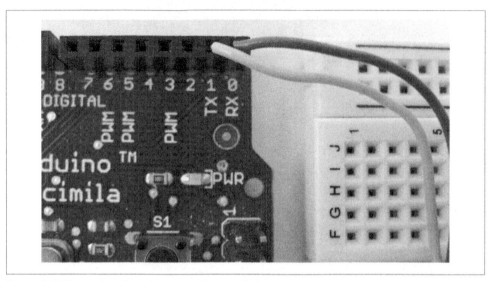

Figure 3-9. Transmit and receive connections on Arduino

Repeat these steps again with the other Arduino and XBee. Figure 3-11 shows the circuit diagram, and Figure 3-12 shows the schematic.

 Sometimes it's a good idea to use a 1 µF capacitor to *decouple* the power supply and filter out high-frequency noise that might interfere with your radio's ability to transmit or receive a clean signal. The Arduino typically provides clean enough power on its own. Decoupling is essential if you use a separate 3.3 V voltage regulator. In that case place the negative leg of the capacitor into ground and the positive leg into power, as near as you can to where your XBee is in the circuit.

Figure 3-10. Power, ground, transmit, and receive connections on XBee. Note that Arduino TX goes to XBee RX and vice versa.

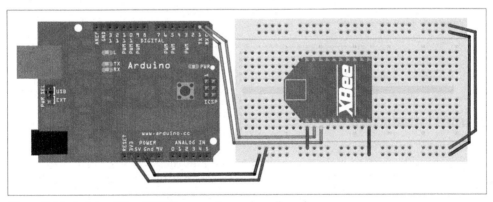

Figure 3-11. Arduino XBee TX/RX connection on breadboard

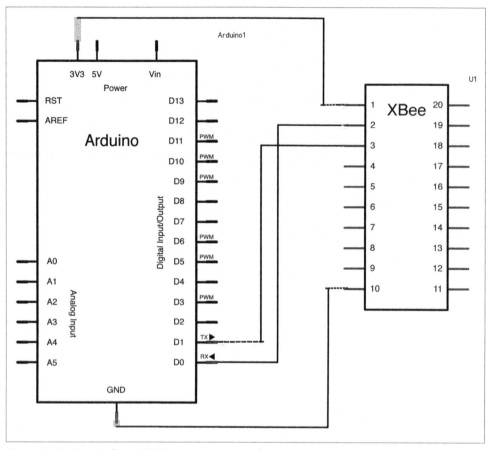

Figure 3-12. XBee Arduino TX/RX connections in schematic view

Doorbell Introduction

Radios aren't much fun on their own, so consider working with a friend on the next two projects, the first to make a simple doorbell and the second to make one with feedback. One of your boards will have the doorbell button input and the other will have a buzzer, speaker, or other noisemaker to act as the doorbell output. The two boards will run different Arduino code, so make sure you load the correct program on each.

Switch Input...

One of your boards will serve as the doorbell button:

1. Pick the board with the coordinator to act as the doorbell button. (Either the board with the coordinator or the one with the router would work equally well, so the choice here is arbitrary.)
2. On that board, attach a momentary switch between power and Arduino digital input 2. Make sure you use a 10K ohm pull-down resistor from digital input 2 to ground. This ensures the pin has a reference to zero voltage while the switch is open.

Figure 3-13 shows the circuit diagram, and Figure 3-14 shows the schematic.

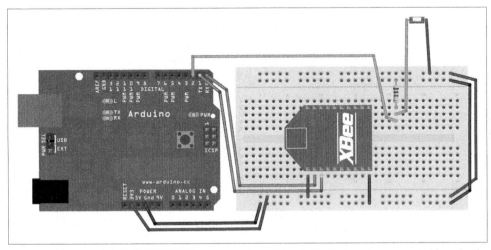

Figure 3-13. Basic doorbell: BUTTON system on breadboard. The button is represented here by the little white thingy, top right.

...and Buzzer Output

Your second board will act as the bell part of the doorbell. On the second board:

1. Attach the red power lead of your buzzer to digital pin 5 of your Arduino board.
2. Attach the black ground lead from your buzzer to ground.
3. If you are already familiar with analog output on the Arduino, you can also use a speaker or other sound output device, in which case employ what you already know to make the proper connections and adjust the Arduino code for that device. Remember that there are a lot of ways to make sound. If you decide to use a servo or to hack a toy, a relay might take the place of the buzzer. Imagination makes this project more fun, so go ahead and amuse your friends or confound your cat.

Figure 3-15 shows the circuit diagram, and Figure 3-16 shows the schematic.

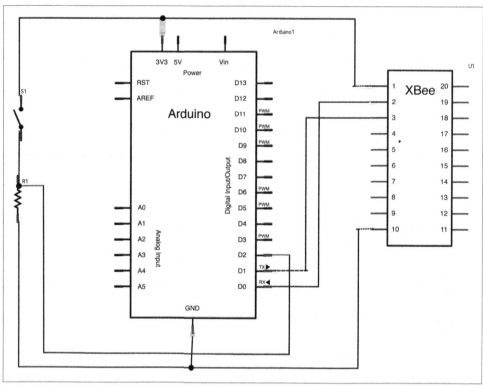

Figure 3-14. Basic doorbell: BUTTON system schematic

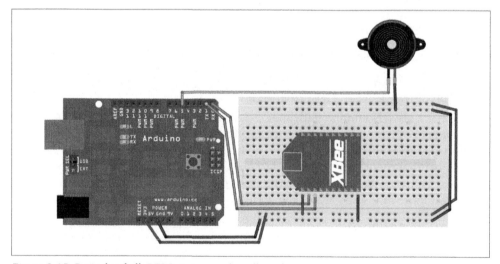

Figure 3-15. Basic doorbell: BELL system on breadboard

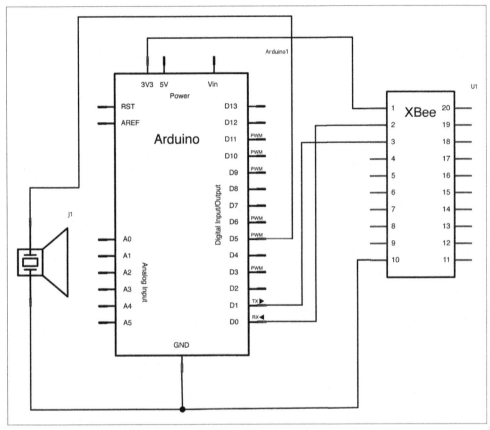

Figure 3-16. Basic doorbell: BELL system schematic

Configure Your XBees

Keep in mind that there are many ways to configure your XBee radios from your computer. We'll be using the CoolTerm (Mac, Windows) terminal program and an XBee Explorer USB adapter. (If you're on Linux, see the sidebar "A Serial Terminal Program for Linux" on page 40 in Chapter 2.) You could also use Digi's X-CTU program and the Digi evaluation board to accomplish the same task, or one of the many other combinations of serial terminal programs and USB adapter setups.

Remember that every XBee radio has a 64-bit serial number address printed on the underside. The beginning or "high" part of the address is 0013A200. The last or "low" part of the address will be different for every radio. It will look something like 4052DAE3.

Write down your coordinator and router radio's addresses so that you can refer to them during configuration:

Coordinator address	Router address
0013A200 _____	0013A200 _____

1. Select the coordinator XBee you labeled with a "C" and place it into the XBee Explorer. (Technically, either radio will work.)

2. Plug the XBee Explorer into your computer.

3. Run the CoolTerm program and press the Options button to configure it.

4. Select the appropriate serial port, which will probably have the words `usbserial` (Mac) or `COM` (Windows) in its name, and check the Local Echo box so you can see your commands as you type them.

5. Click on the Connect button to connect to the serial port.

6. Select a PAN ID between 0x0 and 0xFFFFFFFFFFFFFFFF to define your personal area network.

7. Put the radio into command mode by typing **+++** (remember not to press Return). Next, enter **ATID** followed by the PAN ID you selected. For example, if you selected 0x2001 as your PAN ID, you'd enter **ATID 2001** and press Enter on the keyboard. You should receive OK as a reply. If you don't receive an OK, you probably timed out of command mode and will need to start over with the **+++** and try again.

8. Enter **ATDH** followed by the first "high" part of your radio's *destination* address. In this case we're making a pair of radios, so each one will have the other as its destination. All XBee-brand radios have the same high part—0013A200. Type **ATDH 0013A200** and press Enter on the keyboard.

9. Enter **ATDL** followed by the second "low" part of your radio's *destination* address— the eight-character hexadecimal address of the router radio that follows 0013A200. Type **ATDL** followed by that second part of the destination address, then press Enter on the keyboard. (Don't forget to go into command mode first if you waited more than 10 seconds after your last command.)

10. To save your new settings as the radio's default, type **ATWR** and press Enter.

11. Remove the XBee from the serial adapter.

You set up your second radio in the same way:

1. Select the router XBee you labeled with an "R" and place it into the XBee Explorer.

2. CoolTerm should still be running; if not, repeat steps 2–6 above.

3. Select the *same* PAN ID you entered for your first radio above.

4. Type **+++** to go into command mode. You should receive an OK reply from the radio.

5. Type **ATID** followed by the PAN ID you selected and press Enter on the keyboard. You should receive OK again as a reply.

6. Enter **ATDH** followed by the high part of your radio's *destination* address—always the same for the XBees. Type **ATDH 0013A200** and press Enter on the keyboard. You should receive an OK response.

7. Enter **ATDL** followed by the *low* part of your radio's *destination* address—the eight-character hexadecimal address of the coordinator radio that follows 0013A200. Type **ATDL** followed by that low part of the destination address, then press Enter. You should receive an OK response.

8. Again, save your new settings as the radio's default by typing **ATWR** and pressing Enter.

Your radios are now configured as a pair. Sometimes when people are first starting out with XBees, it takes a few tries to get everything typed in just right to pair the radios. If they don't work at first, don't panic; usually it's just because you missed a single step or made a typo. Try again. Remember that if you are in command mode and type an **AT** command without an argument, the radio will reply with the current setting. This is a good way to check that your configuration is correct.

Program the Arduino Doorbell

 When uploading programs to the Arduino boards, disconnect the wiring from digital pin 0 (RX) first, then reconnect the wiring after loading. If you see an error message from "AVR dude," you probably forgot to do this.

There are two programs that run the doorbell system. One goes with the button and the other goes with the output buzzer or bell. Load this program onto your button board:

```
/*
 * ********* Doorbell Basic BUTTON ********
 * requires pre-paired XBee Radios
 * and the BELL program on the receiving end
 * by Rob Faludi http://faludi.com
 */

#define VERSION "1.00a0"

int BUTTON = 2;


void setup() {
  pinMode(BUTTON, INPUT);
  Serial.begin(9600);
}


void loop() {
  // send a capital D over the serial port if the button is pressed
```

```
      if (digitalRead(BUTTON) == HIGH) {
        Serial.print('D');
        delay(10); // prevents overwhelming the serial port
      }
   }
}
```

Program your *second* board with the bell code. This enables the board to receive a signal from the button board when its switch is activated and to ring the bell:

```
/*
 * ********* Doorbell Basic BELL ********
 * requires pre-paired XBee Radios
 * and the BUTTON program on the receiving end
 * by Rob Faludi http://faludi.com
 */

#define VERSION "1.00a0"

int BELL = 5;


void setup() {
  pinMode(BELL, OUTPUT);
  Serial.begin(9600);
}


void loop() {
  // look for a capital D over the serial port and ring the bell if found
  if (Serial.available() > 0) {
    if (Serial.read() == 'D'){
      //ring the bell briefly
      digitalWrite(BELL, HIGH);
      delay(10);
      digitalWrite(BELL, LOW);
    }
  }
}
```

 Don't forget to reconnect the wiring to digital pin 0 (RX) after loading your code!

Troubleshooting

Sometimes it takes a few tries to get things right. This is a normal part of the learning process, so if your doorbell doesn't ring at first, keep up your good cheer and dig into figuring out the reason why:

1. Start with the simple stuff. Check to make sure your radios are seated correctly in the adapter boards, and that all the wiring is connected properly.

2. Use a serial adapter and a terminal program to check that the radios are paired to communicate with each other. You should be able to use them to chat between

two computers, the way we demonstrated in the previous chapter. Check through the troubleshooting guide for the chat project if it seems like you're having radio issues.

3. If both radios are responding and paired properly, seat them back into the doorbell project. Make sure they're inserted into the sockets properly and facing the correct way.

4. Check your wiring again. The most common problems are simple electrical ones, where one wire is not making good contact or has accidentally been inserted into a socket one away from where it should be. Check for any wires where the bare ends might be touching each other and creating a *short* circuit.

5. Make sure that RX on the Arduino is connected to TX on the XBee and vice versa.

6. Use a multimeter to confirm that your XBee is getting 3.3 V power. Check that both Arduino boards have a power light illuminated.

7. Reload the Arduino code onto both boards. Remember that your button board takes one program, and that your buzzer board takes a different program. Make sure you're loading the right program onto the right Arduino board.

8. If you have any questions about Arduino, the best place to learn more is on the Arduino site itself (*http://arduino.cc*). You'll find a complete reference guide there and extensive forums where you can search for answers or ask a question yourself.

Revelations and Indications

Electricity is invisible. Radio is invisible. How in the world are you going to confirm that your radio has power and is receiving information? There are three places on the XBee radio where you can attach a light to see what's going on:

1. Physical pin 13 is the On/Sleep indicator pin and can tell you if your radio is getting power and currently awake. As you are still a few chapters away from learning how to put the module to sleep, this indicator should always be on. Place an LED with the positive lead (the longer one) into a breadboard socket associated with XBee physical pin 13, and the negative lead (the shorter one) into the ground bus. If the LED lights up, your radio has power and is awake. If it doesn't, check to make sure you put the LED in the right way, with the shorter leg to ground, and that you attached it to the proper XBee pin. After that, check to see that the radio is seated properly in its breakout board, and that it's powered properly with 3.3 volts.

2. An LED placed between the Association indicator on physical pin 15 and ground will light steadily while the radio searches for a network, and then blink once it has associated itself with one. Coordinator radios always blink because they are always associated with the network they created themselves. Router radios (and end devices) will give a steady light when they are powered up and looking for a coordinator. When they find a network to join, their association light will start blinking. If you don't see a light at all, check for power problems. If the light is steady and not blinking, check the configuration of that radio to ensure that it has the same PAN ID as the coordinator and is within range of a radio that it can join.

3. One of the most helpful indicators is on the other side of the XBee, on physical pin 6. Place an LED between this pin and ground. The Received Signal Strength Indicator (RSSI) will light up on this pin when the radio receives information that's addressed to it. By default, the RSSI light will remain on for 10 seconds after it receives information and then go out again. The RSSI LED will be slightly brighter when the signal is strong than when it is weak, but in practice this difference is extremely hard to see. If you think the radio should be receiving information but the RSSI light remains dark, check the configuration of the sender radio to make sure it is on the same PAN ID as the destination, is associated with the network, and has the destination address set correctly.

Figure 3-17 shows an XBee adorned with plenty of LEDs. The bottom right light is the On/Sleep indicator; the top right light is the Association indicator; and the left light shows Received Signal Strength for 10 seconds after data is received. Normally, the On/Sleep light should be on steadily, the Association light should be blinking, and the RSSI light should be on when data is received.

Figure 3-17. XBee with indicator LEDs attached.

Feedback Doorbell

The next project builds on the previous one-way signal to provide two-way feedback that the bell unit has received the doorbell button press and has rung. This is useful so the person at the door knows she actually rang the bell.

Feedback Light

Add an LED as an output from Arduino digital pin 11 on the button board. Figure 3-18 shows the diagram, and Figure 3-19 shows the schematic.

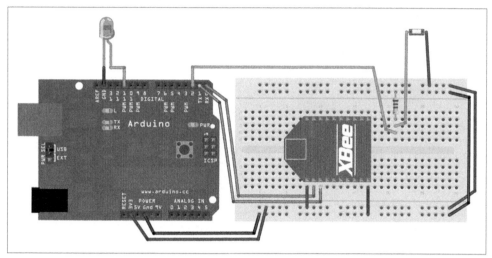

Figure 3-18. Feedback doorbell: BUTTON system on breadboard

Program the Arduino Feedback Doorbell

 Remember to disconnect the wiring from digital pin 0 (RX) first, then reconnect the wiring after loading.

Use the following code for the button board with its new feedback light:

```
/*
 * ********* Doorbell Feedback BUTTON ********
 * requires pre-paired XBee Radios
 * and the BELL program on the receiving end
 * by Rob Faludi http://faludi.com
 */

#define VERSION "1.00a0"

int BUTTON = 2;
int LED = 11;


void setup() {
  pinMode(BUTTON, INPUT);
  pinMode(LED, OUTPUT);
  Serial.begin(9600);
```

```
}

void loop() {
  // send a capital D over the serial port if the button is pressed
  if (digitalRead(BUTTON) == HIGH) {
    Serial.print('D');
    delay(10); // prevents overwhelming the serial port
  }

  // if a capital K is received back, light the feedback LED
  if (Serial.available() > 0 ) {
    if (Serial.read() == 'K') {
      digitalWrite(LED, HIGH);
    }
  }

    // when the button is released, turn off the LED
    if (digitalRead(BUTTON) == LOW) {
      digitalWrite(LED, LOW);
    }

}
```

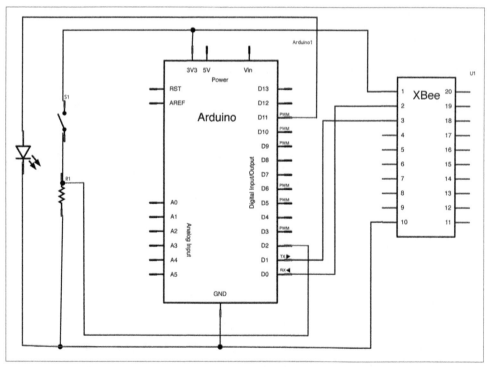

Figure 3-19. Feedback doorbell: BUTTON system schematic

On the second bell board, use this code; it accepts the incoming ring request and responds that the bell has been rung:

```
/*
 * ********* Doorbell Feedback BELL ********
 * requires pre-paired XBee Radios
 * and the BUTTON program on the receiving end
 * by Rob Faludi http://faludi.com
 */

#define VERSION "1.00a0"

int BELL = 5;


void setup() {
  pinMode(BELL, OUTPUT);
  Serial.begin(9600);
}


void loop() {
  // look for a capital D over the serial port and ring the bell if found
  if (Serial.available() > 0) {
    if (Serial.read() == 'D'){
      //send feedback that the message was received
      Serial.print('K');
      //ring the bell briefly
      digitalWrite(BELL, HIGH);
      delay(10);
      digitalWrite(BELL, LOW);
    }
  }
}
```

 Don't forget to reconnect the wiring to digital pin 0 (RX) after loading your code!

Extra: Nap Doorbells and More

There are many ways to take this project one step further. For example, let's imagine a situation where you wanted to take a nap and not be disturbed by the doorbell unless it was extremely urgent. In this case, initial presses of the doorbell button might only illuminate a signal light, rather than waking you with a bell. Eventually, after a large number of presses, the system would kick over into bell mode and wake you up. The caller would initially see a red light at the button to indicate that the bell hadn't been rung yet, then eventually after a large number of presses would see a green light to confirm that the bell had finally rung. Try creating this system or another of your choosing by extending the Feedback Doorbell system with new, useful features. For example, you could create a doorbell that rings only when the button is pressed in a special coded sequence, or a doorbell that can store and replay a history of its rings, or one with an SMS feature to send you a text message when somebody comes calling, or an LCD text display where the visitor can select messages to send with the ring. The possibilities are endless!

CHAPTER 4

Ins and Outs

Congratulations—you now have configurations, communications, and some solid projects under your belt! It's time to take a closer look at the unique features of the XBee brand of ZigBee radio so we can start building fully scalable sensor networks. We begin with input/output concepts and commands, then immediately put these to use in a small set of progressive projects that whimsically inculcate the basics.

The Story of Data

Before getting into the technical aspects of sensing data, it's useful to take a step back and consider why it is we want to collect this type of information in the first place. After all, data has no value by itself. In its purest form, data is just a collection of numbers, and one set of numbers is as good as any other. Our real interest in data always comes from the story it might tell us. Gathering data is the first step in noticing new things in the world, proving a hunch, disproving a fallacy, or teaching a truth. It can also be a path to action. Patterns in data can trigger events, shape public policy, or just determine when it's time to feed the cat. We should always have a purpose in mind when collecting data because that purpose will guide us in how the data is collected. This doesn't mean we need to know what the data will tell us. Our purpose might be to simply gather results to examine for events or patterns that create new questions. This is known in science circles as *exploratory data analysis*—a well-accepted form of initial investigation. In other cases, our plan might be to seek out a highly specific event as a trigger for a fixed response. That sounds complicated, but really it describes most doorbells, including the ones you made in the last chapter. Data is collected from a button for the express purpose of triggering an audio alert. Simple enough, but what else could we learn from it?

Direct, Indirect, Subtext

A huge number of electronic sensors are available. Table 4-1 contains a partial list of those within reach of the average tinkerer.

Table 4-1. Kinds of electronic sensors

Sensor	Detects	Example (SparkFun part numbers unless otherwise noted)
Accelerometer	Accelerations (changes in speed)	SEN-00252
Capacitance	Electrical properties often associated with human touch	SEN-07918
Color	Wavelengths of light	SEN-08663
Flex	angular position and changes	SEN-08606
Force	Physical pressure in an analog scale	SEN-09673
Gas	Alcohol, methane, CO_2, CO, propane, and many others	SEN-08880
		SEN-09404
GSR	Galvanic skin response, typically associated with emotional arousal	http://www.extremenxt.com/gsr.htm
Gyroscope	Rotation	SEN-09423
Hall effect	Magnetic fields	COM-09312
Microphone/acoustic	Sound	BOB-08669
Motion	Changes in relative distance	SEN-08630
Photocell	Light	SEN-09088
Potentiometer	Rotation or linear position on an analog scale	COM-09288
Pressure	Air or fluid pressure	SEN-09694
Pulse	Heartbeat rate	SEN-08660
Ranging	Distance between objects	SEN-00639
Rotary encoder	Rotation on a digital scale	COM-09117
Smoke	Airborne particles	COM-09689
Stretch	Physical deformation or strain	http://www.imagesco.com/sensors/stretch-sensor.html
Switch	Physical pressure on a digital scale	COM-09336
Thermistor	Temperature	SEN-00250
Tilt	Angular attitude	Adafruit 173

Although the table describes detection of one phenomenon per sensor, each sensor is really capable of simultaneously detecting three distinct but intrinsically related categories of events:

Direct or proximal phenomena

These are the incidents that directly trigger the sensor apparatus. For example, in the case of a photocell, the proximal event would be photons striking the sensor. Sometimes the proximal phenomenon is not quite as obvious. For instance, a tilt sensor's proximal trigger would be the repositioning of a metal ball against two electrical contacts. A Hall-effect sensor reports changes in magnetic fields, though that's only rarely the phenomenon of interest.

Indirect or distal phenomena

Distal events are the remote *causes* of the local events actually triggering the sensor. The sun coming out from behind a cloud would be the distal phenomenon that results in a higher reading from a photocell. A window being opened might cause a Hall-effect magnetic sensor to move away from a magnet and open its contacts. These indirect events produce the proximal phenomena that our sensors can respond to, and they are frequently the ones we are most interested in.

Context and subtext

Sometimes neither the proximal or even distal events are what we're after. We aren't interested in magnetic fields at all. In fact, most window openings are not a cause for concern. What we really want to know is if a burglar is entering our house. Our sensor directly detects a change in a magnetic field. That change is an indirect result of a window changing position. But the context is human presence; in this case, definitely a presence that's undesired. Contextual leaps usually entail some degree of uncertainty. A window might swing open in a gust of wind. A houseguest might open up a window that we'd normally leave closed. This creates a need for determining more information to avoid false alerts or missed alarms. Sensing for multiple phenomena can reduce uncertainty. For example, security systems often include window sensors, motion detectors, and pressure mats. When all of these activate simultaneously, it is a more certain indication of criminal presence than hearing from any one on its own.

When choosing a sensor, always think about which category of events you're interested in detecting. Sometimes a surprising relationship can exist where a simple sensor can provide reliable indication of an intricate contextual event. A photocell can report when a bathroom cabinet is opened, by detecting that the interior is no longer totally dark. A microphone can detect the wind noise made when someone blows on a pinwheel, and therefore detect both pressure and presence. A switch on the handle of a toilet might indicate human absence if not triggered for two days, signaling an unsecured front door to lock itself.

Now that we've thought about sensing in theory, let's move on to the practical matter of getting the job done.

I/O Concepts

Each XBee radio has the capability to directly gather sensor data and transmit it, without the use of an external microcontroller. This means that you don't always need something like the Arduino when building simple sensor nodes with XBee radios. In addition, the XBee offers some simple output functions so that basic actuations can also take place without an external microcontroller being present. For example, it's possible to send digital information directly to a standalone XBee radio to have it turn on a light or start up a motor. For clarity, we'll refer to these independent input/output functions as *XBee direct*, to distinguish them from the use of input and output that happens in conjunction with an external microcontroller.

Why XBee Direct?

There are lots of good reasons to use the XBee for direct input or output. By not having an external microcontroller, the overall size of your project is reduced. This is especially important when creating sensors that need to be inconspicuous or fit into tight spaces. By using the XBee alone you'll also save weight, which can be important if the system is to be lofted skyward in a kite or balloon, or worn on your body, or by your pet. When it comes to wearables, lighter is almost always better. Omitting an external microcontroller also reduces power consumption. This can be a critical advantage for projects that run on batteries, a necessary situation for any project that is truly wireless, and something we'll talk about more in Chapter 6. Of course, eliminating the external microcontroller means saving money, and for sensor networks with hundreds of nodes, it can mean saving a *lot* of money. Finally, using the XBee alone is sometimes the least-complicated approach to a project. There's a lot going for the XBee direct model. However, there are also some important trade-offs to consider.

XBee Direct Limitations

Projects that use the XBee alone for its input/output features may face significant limitations compared to projects that incorporate an external microcontroller such as the Arduino. The XBee has limited input and output pins, with no simple way to extend them. Also, the Series 2 hardware that the ZigBee firmware requires doesn't currently support analog output at all, which means it can't be employed to dim a light or control the speed of a motor without additional electronic components. The single biggest limitation is that the basic standalone XBee radio doesn't allow access to any kind of logic. This means no decisions can be made on the local device and no standalone operations can be performed besides transmitting data or changing the state of digital pins as the result of remote commands.

 A new variation of the XBee radio was recently released that incorporates a second microcontroller to allow some forms of local logic. However, this comes at additional cost, will need to be accessed with special programming methods, and requires knowledge of C or Assembly, both lower-level approaches than using Arduino.

XBee I/O Features

The XBee Series 2 hardware offers several flexible features for projects that need simple input and output. There are 10 pins that can be configured either as digital inputs for sensing switches and other things that operate like switches, or as digital outputs for controlling LEDs and small motors directly. Larger loads, including ones that run on alternating current, can be operated using these digital outputs via a relay. The first four of these pins can be configured as analog inputs for sensing a huge array of phenomena that scale over a range, like light, temperature, force, acceleration, humidity, gas levels, and so forth. On the Series 2 radios, there are currently no user-configurable analog or pulse-width modulated (PWM) outputs, so you *cannot* directly control the speed of a motor or the brightness of an LED light. However, the underlying chipset does support these types of outputs so perhaps they will be available in a future firmware upgrade.

XBees have all these different features available, but this doesn't mean you can use them all at once! There are only 10 pins total so you if you have all 10 digital inputs configured, you are out of pins and can't use any digital output or analog input. Happily, the pins can be used in a mix. For example, three analog inputs, four digital inputs, and three digital outputs would be fine. The only other thing to be aware of is that many of the 10 configurable pins are used for other optional duties. These other duties are important in many applications, but they've been carefully selected so that they are ones that don't tend to be needed in remote sensing and actuation projects. For example, some of the duties are serial hardware handshaking (CTS and RTS), an advanced feature that is generally not needed unless there is another microcontroller or logic-based device in the mix. Certain I/O pins do double duty as debugging light outputs for signal strength (RSSI) and association (ASSOC), which are handy for development but generally unimportant on a remote sensor that will not be viewed directly. There are also several pin-controlled sleep features (ON and SLEEP) that are not usually required for stand-alone sensing or actuation. Of course, on the off chance that one or more of those features is required, it would reduce the number of pins available only by one or two, so you'll generally have enough left over to cover the vast majority of application projects you can dream up. Table 4-2 shows the input/output pin names with physical numbers, corresponding AT commands, and other functions. Note that DIO8 and DIO9 are not supported in the current firmware so they can't be used for I/O at this time. Figure 4-1 shows the I/O pins on a breakout board.

Table 4-2. Input/output pin names with physical numbers, commands, and other functions

Pin name	Physical pin #	AT command	Other functions
DIO0, AD0	20	D0	Analog input, Commissioning Button
DIO1, AD1	19	D1	Analog input
DIO2, AD2	18	D2	Analog input
DIO3, AD3	17	D3	Analog input

Pin name	Physical pin #	AT command	Other functions
DIO4	11	D4	
DIO5	15	D5	Association indicator
DIO6	16	D6	RTS
DIO7	12	D7	CTS
(DIO8)	9	None	Pin sleep control, DTR
(DIO9)	13	None	On/Sleep indicator
DIO10	6	P0	Received Signal Strength Indicator (RSSI)
DIO11	7	P1	
DIO12	4	P2	

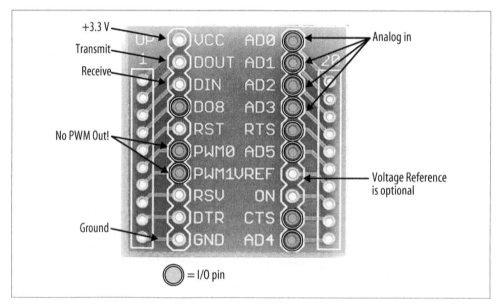

Figure 4-1. I/O pins as seen on a breakout board

AT Configuration I/O Commands

To configure the XBee radio for direct input, output, or both, you'll use a set of AT commands that select each pin's mode and the sample rate for sending the data. There are several steps involved in getting this done, so read carefully through this section at least once before starting to configure your radio.

Here's the basic I/O command set:

ATD0...ATD7

Configures pins 0 through 7 for I/O mode (pins 8 and 9 are not supported in the current firmware version). The number after the D indicates which pin you'll be configuring. The command is followed by a numeric code that indicates whether the pin is tasked with digital input, output, analog input (pins 0 to 3 only), some other function, or nothing at all. For example, to configure I/O pin 2 as a digital input (code 3), the command would be ATD23. See the I/O settings codes in Table 4-3 for a complete list of the codes.

ATP0...ATP2

Configures pins 10 11, and 12 for I/O mode (there's a P3 for pin 13, but it is not supported in the current firmware). Again, the number after the P indicates which pin you'll be configuring, and is followed by a numeric code to indicate what purpose the pin will serve—digital in, digital out, or nothing. For example, to configure I/O pin 11 as a high digital output (code 5) the command would be ATP15. Pins 10-12 do not support any analog functions.

ATIR

This sets the I/O sample rate—how frequently to report the current pin state and transmit it to the destination address. The rate is set in milliseconds, using *hexadecimal* notation. So, for example, let's say you want to take a sample 10 times every second. There are 1,000 milliseconds in a second so we divide this by 10 to get 100 milliseconds. Now we just need to find the hexadecimal equivalent of 100. This happens to be 0x64, so the command would be ATIR64. To disable periodic sampling, simply set ATIR to zero.

ATWR

Don't forget to write the configuration to firmware using ATWR so that the next time your radio powers up it retains the correct settings!

The settings codes for each I/O pin (Table 4-3) designate whether it will do nothing, perform a built-in function, take analog input, take digital input, or give digital output.

Table 4-3. I/O settings codes for use with ATDx and ATPx (where x is the pin #)

ATDx or ATPx followed by:	Purpose:
0	Disables I/O on that pin
1	Built-in function, if available on that pin
2	Analog input, only on pins D0 through D3
3	Digital input
4	Digital output, low (0 volts)
5	Digital output, high (3.3 volts)

Analog input pins D0 through D3 read a range from 0 volts to 1.2 volts maximum. Voltages above 1.2 are ignored and result in the same maximum reading. Because most circuits using the XBee Series 2 run at 3.3 volts, if your input is a variable resistor, like a photoresistor, flex sensor, or force sensor, you'll need to create a voltage divider circuit that cuts maximum voltage by two-thirds to keep it within the range of the analog-digital converter (ADC).

The formula for voltage divider output between the two resistors is:

$$V_{out} = \frac{R_2}{R_1 + R_2} \times V_{in}$$

A fast implementation for transforming a 3.3 V input into one that stays below 1.2 V max is to have the fixed resistor R_1 be twice the maximum resistance of the variable resistor R_2. So in the circuit shown in Figure 4-2, if R_2 is a flex sensor with a maximum resistance of 10K ohm, then R_1 would be a 20K ohm fixed resistor. Or, for a photocell rated at 300 ohms, a good choice of fixed resistor would be 600 ohms.

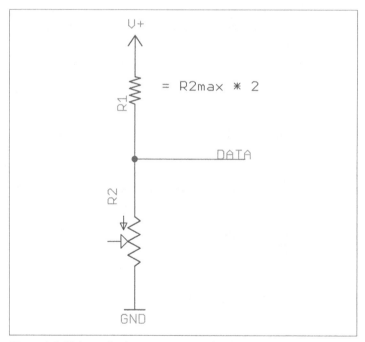

Figure 4-2. Voltage divider circuit to map 3.3 V range to 1.2 V range

Advanced I/O Commands

Several other AT commands may come in handy for projects with special I/O needs. These are worth knowing about even if you don't need to use them right away. The XBee manual has detailed specifications for each of these commands:

AT%V

> Returns the current supply voltage for the module. This is useful for keeping track of battery status.

ATPR

> Configures the internal 30K ohm pull-up resistors, using a binary value to set for each pin you've configured as an input. This is useful if your input component is a momentary digital switch that connects to ground, so you don't need to add the required external pull-up resistor. By default, the internal pull-ups are all enabled.

ATIC

> Configures the digital I/O pins to monitor for changes in state, using a binary value to set for each pin. The pin(s) would also need to be configured as digital inputs. When change-detection is enabled, a sample is sent immediately any time a pin shifts from low to high or vice versa. This is useful if you are monitoring a switch, and care about triggering a transmission only when a button is pressed or released.

Romantic Lighting Sensor

Wireless networking is not nearly as tricky as navigating romance. Luckily, the former can help you with the latter, as this next project will demonstrate. Imagine for a moment that you are a brilliant engineer, hacker, interaction designer, or scientist—and perhaps you actually are. Let's say you've mastered math, manual skills, and usability, but nothing in your schooling has prepared you for the daunting task of setting a scene where love can blossom. What to do? The dining table is laid out perfectly; your date is moments away from ringing your wireless doorbell; now how to set the lights? We all know that glaringly bright lighting tends to hamper courtship. This is a date after all, not an interrogation. On the other hand, dimming the lights too far can seem creepy. What you need is a sensing system that lets you know you've lit things in the sweet spot for romance.

To get you started, here's a project that creates a remote wireless lighting sensor with a base station that lights a green LED when the mood is just right. It also happens to be a fine example for developing a variety of your own wireless I/O projects.

Basic Romantic Lighting Sensor

We'll start by creating a simple wireless lighting sensor that gives feedback at the base station.

Parts

- Two solderless breadboards (MS MKKN2, AF 64, DK 438-1045-ND, SFE PRT-09567)
- Hookup wire or jumper wire kit (MS MKSEEED3, AF 153, DK 923351-ND, SFE PRT-00124)
- One Arduino board (MS MKSP4, SFE DEV-09950, AF 50)
- USB A-to-B cable for Arduino (AF 62, DK 88732-9002, SFE CAB-00512)
- Two AA battery holders with connection wires (RS 270-408, SFE PRT-09547)
- Two AA batteries (alkaline or NIMH rechargeable, fully charged) (RS 23-873, SFE PRT-09100 or PRT-00335)
- Two 5 mm LEDs (DK 160-1707-ND, RS 276-041, SFE COM-09590)
- One 20K ohm resistor (or twice the max value of your photoresistor) (DK P20KBACT-ND, SFE COM-08374 * 2 in series)
- One 10K ohm photoresistor (also called an *LDR* or light-dependent resistor) (AF 161, DK PDV-P8001-ND, SFE SEN-09088)
- One XBee radio (Series 2/ZB firmware) configured as a ZigBee Coordinator API mode (Digi: XB24-Z7WIT-004, DK 602-1098-ND)
- One XBee radio (Series 2/ZB firmware) configured as a ZigBee Router AT mode (Digi: XB24-Z7WIT-004, DK 602-1098-ND)
- Two XBee breakout boards with male headers and 2 mm female headers installed (AF 126 [add SFE PRT-00116], SFE BOB-08276, PRT-08272, and PRT-00116)
- XBee USB Serial adapter (XBee Explorer, Digi Evaluation board, or similar) (AF 247, SFE WRL-08687)
- USB cable for XBee adapter (AF 260, SFE CAB-00598)
- Wire strippers (AF 147, DK PAL70057-ND, SFE TOL-08696)

Prepare your coordinator radio

Write down your coordinator and router radios' addresses (printed on the back) so that you can refer to them during configuration:

Coordinator address	Router address
0013A200 _____	0013A200 _____

1. Follow the instructions under "Reading Current Firmware and Configuration" on page 35 in Chapter 2 to configure one of your radios as a ZigBee Coordinator API.

Your *coordinator* radio *must* use the API firmware for this project to work, because I/O data is delivered only in API mode. Be sure to select the API version for your coordinator!

When you change from AT to API mode using X-CTU, you may get an error message that the radio is no longer communicating. Go back to the PC Settings tab and check the Enable API box (Figure 4-3) to enable communications with your radio.

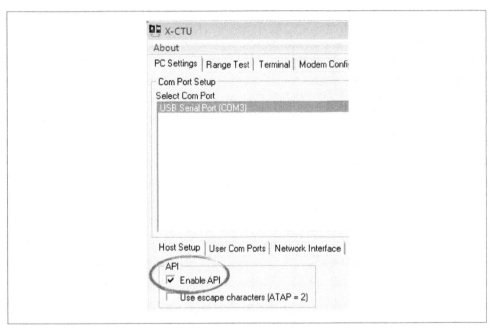

Figure 4-3. Enable API checkbox on PC Settings tab in X-CTU

2. Once a radio has been set to API mode, it can *only* be configured in X-CTU. You will not be able to make adjustments to this radio's configuration in CoolTerm or any other terminal program. Use X-CTU to configure the coordinator with a PAN ID (between 0x0 and 0xFFFFFFFFFFFFFFFF) you've selected. Write down this PAN ID so you can program your router radio with the same one. Every radio in your network must use the same PAN ID so they can communicate with each other:

Pan ID:

3. Use X-CTU (Figure 4-4) to set ATDH to the high part of your *router* radio's address (always 0013A200 for XBees) and ATDL to the remainder of your *router* radio's address (the unique part of the number you noted above).

4. Click on the Write button to save your settings to the radio.

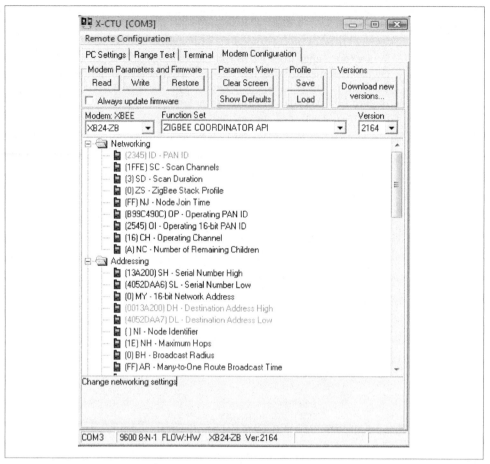

Figure 4-4. Setting ID, DH, and DL in X-CTU

Prepare your router radio

1. Follow the instructions under "Reading Current Firmware and Configuration" on page 35 in Chapter 2 to configure one of your radios as a ZigBee Router AT.

> Your *router* radio will use the *AT* firmware, so you can easily configure it using a serial terminal. Be sure you select the AT version for your router!
>
> When you change from an API radio to an AT radio, you may get an error message that the radio is no longer communicating. If so, go back to the PC Settings tab and *un*check the Enable API Mode box (Figure 4-5).

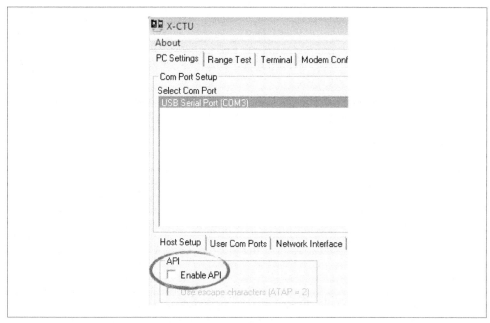

Figure 4-5. Disabled API checkbox on PC Settings tab in X-CTU

2. Label the coordinator radio with a "C" so you know which one it is later on. Label the router radio with an "R."

Prepare the Sensor Board

We'll use the CoolTerm terminal program (Mac, Windows) and an XBee Explorer USB adapter again to set up your radios. (If you're on Linux, see the sidebar "A Serial Terminal Program for Linux" on page 40 in Chapter 2.)

Configure your router XBee

1. Select the router XBee you labeled with an "R" and place it into the XBee Explorer.
2. Plug the XBee Explorer into your computer.
3. Run the CoolTerm program and press the Options button to configure it.
4. Select the appropriate serial port, and check the Local Echo box so you can see your commands as you type them.
5. Click on the Connect button to connect to the serial port.
6. Type **+++** to go into command mode. You should receive an OK reply from the radio.
7. Select the **same** PAN ID you entered for your first radio. PAN ID: _____

8. Type **ATID** followed by the PAN ID you selected and press Enter on the keyboard. You should receive OK again as a reply.

9. Enter **ATDH** followed by the *high* part of your radio's *destination* address—always the same for the XBees. Type **ATDH 0013A200** and press Enter on the keyboard. You should receive an OK response.

10. Enter **ATDL** followed by the *low* part of your radio's *destination* address—the eight-character hexadecimal address of the coordinator radio that follows 0013A200. Type **ATDL** followed by that low part of the destination address, then press Enter. You should receive an OK response. Remember that your destination will be the coordinator radio.

11. Enter **ATJV1** to ensure that your router attempts to rejoin the coordinator on startup.

12. Enter **ATD02** to put pin 0 in analog mode.

13. Enter **ATIR64** to set the sample rate to 100 milliseconds (hex 64).

14. Save your new settings as the radio's default by typing **ATWR** and pressing Enter.

 It's not a bad idea to check your configurations after you enter them. For example, to check that you entered the destination address correctly, from command mode type **ATDL** and press Return to see the current setting.

Connect power from battery to breadboard

Your remote sensor will use a breadboard connected to two AA batteries:

1. Hook up the positive (usually red) battery lead to one of the power rails on the breadboard.

2. Hook up the ground (usually black) battery lead to a ground rail on the breadboard.

3. Hook up power and ground across the breadboard so that the rails on both sides are live.

Router XBee connection to battery

1. With the *router* XBee mounted on its breakout board, position the breakout board in the center of your other breadboard so that the two rows of male header pins are inserted on opposite sides of the center trough.

2. Use red hookup wire to connect pin 1 (VCC) of the XBee to 3-volt battery power.

3. Use black hookup wire to connect pin 10 (GND) of the XBee to ground.

Photoresistor input

The battery-powered board with the router radio will be your remote sensor. On that board:

1. Attach a photoresistor between ground and XBee digital input 0 (physical pin 20).
2. Make sure you use the 20K ohm (or other value that's double your photoresistor's max value) pull-up resistor from digital input 0 to power. This ensures the sensor has a proper voltage divider circuit, which is required to get correct readings.

Figure 4-6 shows the layout of the board, and Figure 4-7 shows the schematic.

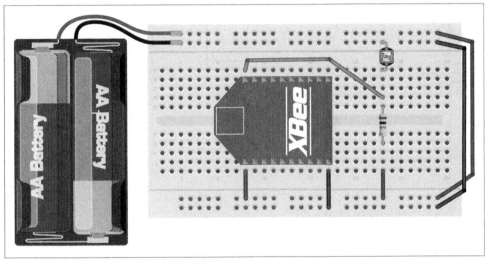

Figure 4-6. Romantic lighting sensor BASIC SENSOR breadboard layout

Prepare the Base Station

Your base station radio will use a breadboard connected to an Arduino board.

Connect power from Arduino to breadboard

1. Hook up a red wire from the 3.3 V output of the Arduino to one of the power rails on the breadboard.
2. Hook up a black wire from either ground (GND) connection on the Arduino to a ground rail on the breadboard.
3. Hook up power and ground across the breadboard so that the rails on both sides are live.

 Make sure you are using 3.3 V power.

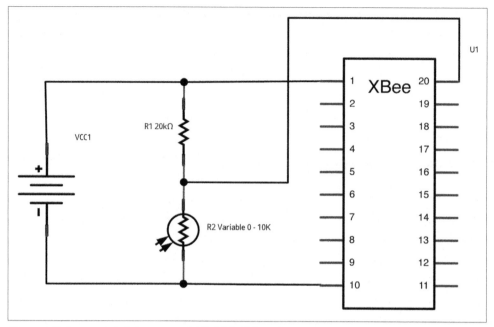

Figure 4-7. Romantic lighting sensor BASIC SENSOR schematic

Coordinator XBee connection to Arduino

1. With the *coordinator* XBee mounted on its breakout board, position the breakout board in the center of one of your breadboards so that the two rows of male header pins are inserted on opposite sides of the center trough.

2. Use red hookup wire to connect pin 1 (VCC) of the XBee to 3.3-volt power.

3. Use black hookup wire to connect pin 10 (GND) of the XBee to ground.

4. Use yellow (or another color) hookup wire to connect pin 2 (TX/DOUT) of the XBee to digital pin 0 (RX) on your Arduino.

5. Finally, use blue (or another color) hookup wire to connect pin 3 (RX/DIN) of your XBee to digital pin 1 (TX) on your Arduino.

Light output

1. Attach the positive (longer) lead of an LED to Arduino digital pin 11.

2. Attach the shorter ground lead from your LED to ground.

3. If you prefer to use another output, like an audio buzzer or pager motor, you can hook it up in the same way. Perhaps your romance chops are best demonstrated by a puff of scented air freshener. Then again, maybe a monkey playing the drums is more your style. The key to romance is being yourself, so don't hesitate to get creative!

Figure 4-8 shows the layout of the board, and Figure 4-9 shows the schematic.

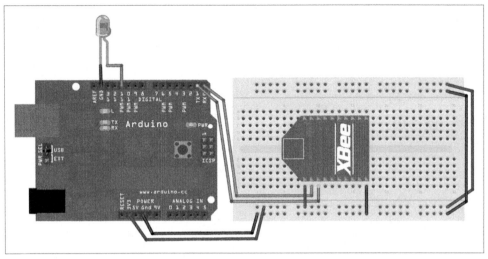

Figure 4-8. Romantic lighting sensor BASE breadboard configuration

Program the romantic lighting sensor base station

The romantic lighting sensor base station uses the following Arduino program. Upload it to your Arduino board and you're ready to test the mood:

 When uploading programs to the Arduino board, disconnect the wiring from digital pin 0 (RX) first, then reconnect the wiring after loading.

```
/*
 * *********ROMANTIC LIGHTING SENSOR ********
 * detects whether your lighting is
 * setting the right mood
 * USES PREVIOUSLY PAIRED XBEE ZB RADIOS
 * by Rob Faludi http://faludi.com
 */

/*
*** CONFIGURATION ***

SENDER: (REMOTE SENSOR RADIO)
ATID3456 (PAN ID)
ATDH -> set to SH of partner radio
ATDL  -> set to SL of partner radio
ATJV1 -> rejoin with coordinator on startup
ATD02  pin 0 in analog in mode
ATIR64 sample rate 100 millisecs (hex 64)
```

```
   * THE LOCAL RADIO _MUST_ BE IN API MODE *

  RECEIVER: (LOCAL RADIO)
  ATID3456 (PAN ID)
  ATDH -> set to SH of partner radio
  ATDL  -> set to SL of partner radio

  */

#define VERSION "1.02"

int LED = 11;
int debugLED = 13;
int analogValue = 0;


void setup() {
  pinMode(LED,OUTPUT);
  pinMode(debugLED,OUTPUT);
  Serial.begin(9600);
}


void loop() {
  // make sure everything we need is in the buffer
  if (Serial.available() >= 21) {
    // look for the start byte
    if (Serial.read() == 0x7E) {
      //blink debug LED to indicate when data is received
      digitalWrite(debugLED, HIGH);
      delay(10);
      digitalWrite(debugLED, LOW);
      // read the variables that we're not using out of the buffer
      for (int i = 0; i<18; i++) {
        byte discard = Serial.read();
      }
      int analogHigh = Serial.read();
      int analogLow = Serial.read();
      analogValue =  analogLow + (analogHigh * 256);
    }
  }

  /*
   * The values in this section will probably
   * need to be adjusted according to your
   * photoresistor, ambient lighting, and tastes.
   * For example, if you find that the darkness
   * threshold is too dim, change the 350 value
   * to a larger number.
   */

  // darkness is too creepy for romance
  if (analogValue > 0 && analogValue <= 350) {
    digitalWrite(LED, LOW);
  }
```

```
  // medium light is the perfect mood for romance
  if (analogValue > 350 && analogValue <= 750) {
    digitalWrite(LED, HIGH);
  }
  // bright light kills the romantic mood
  if (analogValue > 750 && analogValue <= 1023) {
    digitalWrite(LED, LOW);
  }

}
```

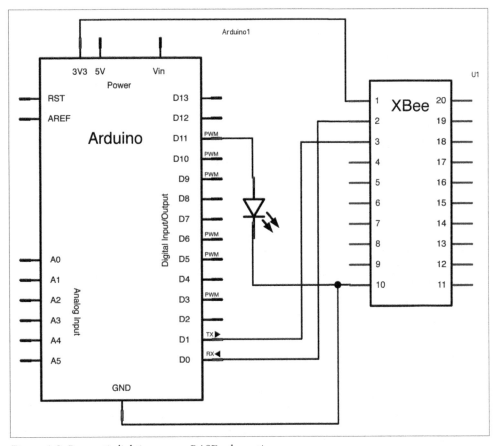

Figure 4-9. Romantic lighting sensor BASE schematic

Troubleshooting

If things don't work at first, here are some steps to take to try to figure out what's wrong:

1. Check all your electrical connections to make sure there are no loose wires and that all the components are connected properly.

2. Check the coordinator configuration in X-CTU again, including that the correct modem type (XB24-ZB) and function set (ZigBee Coordinator API) have been selected. Also check that the PAN ID, destination high, and destination low are configured as you expect. Remember the destination is the *other* radio.

3. Check the router configuration in X-CTU to confirm that the correct modem type (XB24-ZB) and function set (ZigBee Router AT) have been selected. Also check that the PAN ID, destination high, and destination low are configured as you expect, and that ATJV, ATD0, and ATIR have been configured as described above.

4. Make sure that the Arduino is programmed with the correct code for this project (the basic version above and the feedback version below have different code and must be matched to the correct board setup and radio settings).

5. The debug LED on the Arduino board (pin 13) will flash if you are receiving data. If this light is flashing but your output light doesn't change, try adjusting the sensor threshold values in the Arduino code.

6. An LED placed from the ASSOC pin of the XBee (physical pin 15) to ground should show a flashing light.

7. An LED placed from the RSSI pin of the XBee (physical pin 6) to ground should show a steady light when the radio is receiving information. If messages stop coming in, this light will time out and go dark after 10 seconds.

8. Use a multimeter to see if the voltage at the D0 pin of the XBee (physical pin 20) varies with changes in the lighting. It should be somewhere in the range between 0 and 1.2 volts and change as you shadow the light sensor with your hand.

9. We are not always able to see our own mistakes. Have a friend check everything for you. Sometimes only a second pair of eyes will catch the one (or more) issues standing in the way of success.

10. When all else fails: Try taking a break and coming back to the project after a good night's rest. Many of midnight's intractable puzzles are morning's simple fix.

Romantic Lighting Sensor with Feedback

The basic sensor works pretty well as long as you are at the base station. However, it's a pain to run back and forth between the sensor and the base to see if the mood is right. Let's improve on things by putting the feedback right where the sensor is. This is also a nice example to start with for any project where you want both sensing and actuation on a remote device.

Add light output to the sensor

On the sensor board:

1. Attach the positive (longer) lead of an LED to XBee digital input 1 (physical pin 19).

2. Attach the shorter ground lead from your LED to ground.

Figure 4-10 shows the layout of the board, and Figure 4-11 shows the schematic.

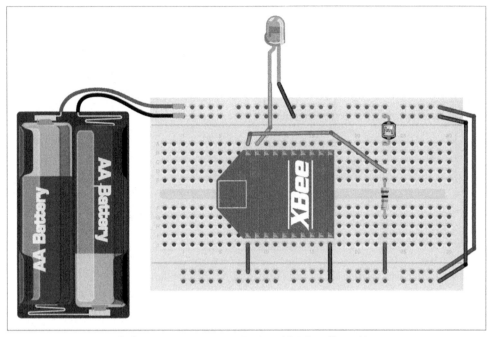

Figure 4-10. Romantic lighting sensor FEEDBACK SENSOR breadboard layout

Configure your router XBee

We'll use the CoolTerm terminal program and an XBee Explorer USB adapter again to set up the radios. The setup is the same as in the basic version above, with the addition of a digital output pin to control the sensor LED:

1. Select the router XBee you labeled with an "R" and place it into the XBee Explorer.
2. Plug the XBee Explorer into your computer.
3. Run the CoolTerm program and press the Options button to configure it.
4. Select the appropriate serial port, and check the Local Echo box so you can see your commands as you type them.
5. Click on the Connect button to connect to the serial port.
6. Type **+++** to go into command mode. You should receive an OK reply from the radio.
7. Enter **ATD14** to put pin 1 in low digital output mode.
8. Save your new settings as the radio's default by typing **ATWR** and pressing Enter.

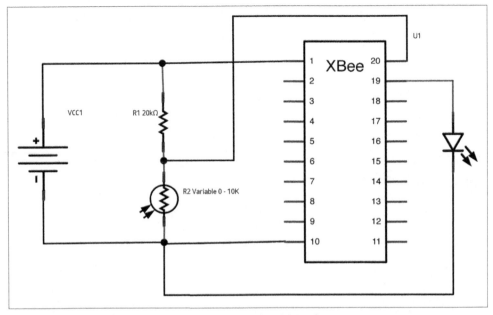

Figure 4-11. Romantic lighting sensor FEEDBACK SENSOR schematic

Program the romantic lighting sensor with feedback base station

The romantic lighting sensor with feedback base station uses the following Arduino program. Upload it to your Arduino board and you're ready to test the mood right from the sensor:

 When uploading programs to the Arduino board, disconnect the wiring from digital pin 0 (RX) first, then reconnect the wiring after loading.

```
/*
 * ********ROMANTIC LIGHTING SENSOR WITH FEEDBACK********
 * detects whether your lighting is
 * setting the right mood and shows
 * you the results on the sensor module
 * USES PREVIOUSLY PAIRED XBEE ZB RADIOS
 * by Rob Faludi http://faludi.com
 */

/*
*** CONFIGURATION ***

SENDER: (REMOTE SENSOR RADIO)
ATID3456 (PAN ID)
ATDH -> set to SH of partner radio
ATDL  -> set to SL of partner radio
ATJV1 -> rejoin with coordinator on startup
ATD02  pin 0 in analog in mode with a photo resistor
```

```
              (don't forget the voltage divider circuit--resistor
              to ground is good)
    ATD14  pin 1 in digital output (default low) mode with an
              LED from that pin to ground
    ATIR64 sample rate 100 millisecs (hex 64)


    * THE LOCAL RADIO _MUST_ BE IN API MODE *

    RECEIVER: (LOCAL RADIO)
    ATID3456 (PAN ID)
    ATDH -> set to SH of partner radio
    ATDL  -> set to SL of partner radio

    */

#define VERSION "1.02"

int LED = 11;
int debugLED = 13;
int analogValue = 0;
int remoteIndicator = false;      // keeps track of the desired remote
                                  // on/off state
int lastRemoteIndicator = false; // record of prior remote state
unsigned long lastSent = 0;       // records last time the remote was
                                  // reset to keep it in sync


void setup() {
  pinMode(LED,OUTPUT);
  pinMode(debugLED,OUTPUT);
  Serial.begin(9600);
}


void loop() {
  // make sure everything we need is in the buffer
  if (Serial.available() >= 23) {
    // look for the start byte
    if (Serial.read() == 0x7E) {
      //blink debug LED to indicate when data is received
      digitalWrite(debugLED, HIGH);
      delay(10);
      digitalWrite(debugLED, LOW);
      // read the variables that we're not using out of the buffer
      // (includes two more for the digital pin report)
      for (int i = 0; i<20; i++) {
        byte discard = Serial.read();
      }
      int analogHigh = Serial.read();
      int analogLow = Serial.read();
      analogValue =  analogLow + (analogHigh * 256);
    }
  }
```

```
/*
 * The values in this section will probably
 * need to be adjusted according to your
 * photoresistor, ambient lighting, and tastes.
 * For example, if you find that the darkness
 * threshold is too dim, change the 350 value
 * to a larger number.
 */

// darkness is too creepy for romance
if (analogValue > 0 && analogValue <= 350) {
  digitalWrite(LED, LOW);
  remoteIndicator = false;
}
// medium light is the perfect mood for romance
if (analogValue > 350 && analogValue <= 750) {
  digitalWrite(LED, HIGH);
  remoteIndicator = true;
}
// bright light kills the romantic mood
if (analogValue > 750 && analogValue <= 1023) {
  digitalWrite(LED, LOW);
  remoteIndicator = false;
}

// set the indicator immediately when there's a state change
if (remoteIndicator != lastRemoteIndicator) {
  if (remoteIndicator==false) setRemoteState(0x4);
  if (remoteIndicator==true) setRemoteState(0x5);
  lastRemoteIndicator = remoteIndicator;
}

// reset the indicator occasionally in case it's out of sync
if (millis() - lastSent > 10000 ) {
  if (remoteIndicator==false) setRemoteState(0x4);
  if (remoteIndicator==true) setRemoteState(0x5);
  lastSent = millis();
}


}

void setRemoteState(int value) {  // pass either a 0x4 or 0x5 to turn the pin on/off
  Serial.print(0x7E, BYTE); // start byte
  Serial.print(0x0, BYTE);  // high part of length (always zero)
  Serial.print(0x10, BYTE); // low part of length (the number of bytes
                            // that follow, not including checksum)
  Serial.print(0x17, BYTE); // 0x17 is a remote AT command
  Serial.print(0x0, BYTE);  // frame id set to zero for no reply
  // ID of recipient, or use 0xFFFF for broadcast
  Serial.print(00, BYTE);
  Serial.print(00, BYTE);
  Serial.print(00, BYTE);
  Serial.print(00, BYTE);
  Serial.print(00, BYTE);
```

```
    Serial.print(00, BYTE);
    Serial.print(0xFF, BYTE); // 0xFF for broadcast
    Serial.print(0xFF, BYTE); // 0xFF for broadcast
    // 16 bit of recipient or 0xFFFE if unknown
    Serial.print(0xFF, BYTE);
    Serial.print(0xFE, BYTE);
    Serial.print(0x02, BYTE); // 0x02 to apply changes immediately on remote
    // command name in ASCII characters
    Serial.print('D', BYTE);
    Serial.print('1', BYTE);
    // command data in as many bytes as needed
    Serial.print(value, BYTE);
    // checksum is all bytes after length bytes
    long sum = 0x17 + 0xFF + 0xFF + 0xFF + 0xFE + 0x02 + 'D' + '1' + value;
    Serial.print( 0xFF - ( sum & 0xFF) , BYTE ); // calculate the proper checksum
    delay(10); // safety pause to avoid overwhelming the
               // serial port (if this function is not implemented properly)
}
```

API Ahead

These last code examples contain something we haven't really looked at yet, API mode. The next chapter will explore the XBee Application Programming Interface, a highly structured way of communicating with your XBee radio. You've already used it, so let's find out how it works and why it is essential to certain projects.

CHAPTER 5

API and a Sensor Network

Here the plot heats up. You now have everything you need to conquer the XBee's *application programming interface*. This is something we need to do so we can use all the data our networks can provide. We will start with simple concepts and scaffold you up to a full understanding of the structured API communication frames. That will get you ready to create a fully scalable sensor network of your own, using the example at the end of the chapter.

What's an API?

An application programming interface (API) is simply a set of standard interfaces created to allow one software program to interact with another. APIs let one computer application request services from another application in a standard manner. For our purposes, the most important thing to note is that APIs are specifically engineered to enable computers to talk efficiently to other computers. They are not generally designed for direct human interaction.

So far, we've been using the XBee radios in transparent/command mode. For example, in the simple chat we set up in Chapter 2, we were able to type text at a keyboard to enter command mode, then issue AT commands by typing them right in. When we were done with configuration, we exited command mode and went right into transparent mode, where everything typed at the keyboard was transferred verbatim to the destination radio and read directly on the screen. This was a simple way to get started with wireless networking, and it's one of the great strengths of the XBee platform. It's very easy for humans to get started using direct interactions in the transparent/command modes. However, there is also a catch. When interactions are made easy for humans, they are not as robust, explicit, and efficient for computers. Computers care about things like algorithmic error correction, airtight mode identification, and efficient data transfer to get their job done quickly, predictably, and reliably. At the same time, they could care less about readability. Computers prefer to deal with numbers, and do best when the organization of these numbers provides an unambiguous and highly structured method for transfer. This is where API mode comes in to save the day. By providing an interface for programmatic communication with the XBee, API mode enables the radios to serve humans and computers equally well, each according to their needs.

 Transparent, command, and API interaction modes with an XBee are *local* to that particular radio. "Local" means that they apply to interactions with users, computers, or microcontrollers that take place via the XBee's serial connection (also known as its UART). Wireless communications between XBees are independent of the local interaction mode, as shown in Figure 5-1. So a radio in transparent mode can send to another in API mode just fine. It's only in local serial communications where transparent/command mode and API mode make a difference.

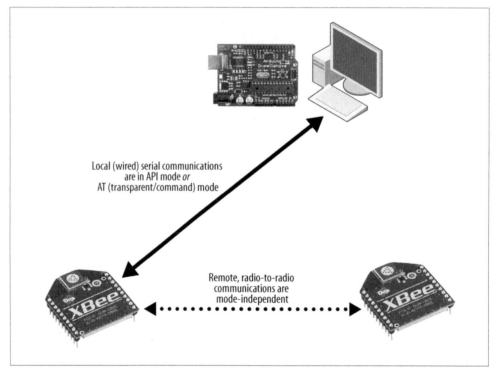

Local (wired) serial communications
are in API mode *or*
AT (transparent/command) mode

Remote, radio-to-radio
communications are
mode-independent

Figure 5-1. *Local communications over serial wires happen in API mode or AT (transparent/ command) mode. Wireless communications are not affected by the choice of local modes.*

Before we dive into the XBee API, let's review protocols in general and build a hypothetical one to examine just how they work.

Protocols

Every transfer of information requires a protocol. Protocols are easy to understand; they're simply agreements on some rules for communication. There are established protocols for wireless computer communications, just as there are protocols for two human beings who want to have a casual conversation. Both people and computers face the same types of communication problems and we solve them in very similar ways.

Humans

Let's say that Jane and Michael have something to discuss. If Michael starts off in nonstop Chinese and Jane begins to shout over him in Icelandic, very little is going to be accomplished. It's best for them to agree on a common language first, and then follow some rules for speaking and listening so that an exchange of information can take place. If Jane begins speaking first, Michael will wait for her to finish, and then respond to her remarks. While Michael is talking, Jane will listen. If a fire truck goes by and Jane can't hear, she'll ask Michael to repeat whatever she missed. If what Michael says doesn't make sense to Jane, she'll either ignore his misspeak or request clarification depending upon the specifics of that particular statement. Much of the protocol of human conversation is so well ingrained that we usually don't think about the rules. We just follow them naturally. It's only when we discard protocol, by talking over each other, mumbling incoherently, or failing to voice our confusion that communications fall apart.

Computers

When computers talk to each other, they try to fulfill a given purpose in the simplest manner possible. In some contexts, that can be pretty simple indeed! For example, let's look at the least complicated communications protocol: basic streaming. One computer talks nonstop and the other listens. This is the perfect solution for conveying simple data from one point to another as long as some errors can be tolerated. More complicated protocols will define whether there's some kind of handshaking involved to set up the exchange, timing issues, what replies are sent in response to what messages, routing strategies, and so forth. But we don't need to worry about any of that for now because we're keeping it simple.

Let's say we want to send a number between 0 and 255 to represent in real time how bright it is outside.

We use the range 0–255 because 255 is the largest number that can be represented in a single byte of data. All common forms of serial communication break data up into *bytes*. A byte is a set of eight digital *bits*. A single bit can be either 0 or 1, thereby representing two states. Add another bit and you now have four states: 00, 01, 10, and 11. A third bit allows for eight states (000, 001, 010, 011, 100, 101, 110, and 111), and so on and so forth until you get to eight bits that can represent 256 different states (including the zero state). That's a byte! In decimal, the numbers go from zero to 255, and in hexadecimal notation they go from 0x0 to 0xFF. For more on binary and bytes, see: *http://en.wikipedia .org/wiki/Byte*.

If we send our brightness data once every second, it would look something like this:

    136...137...137...138...138...138...139...135...128...110...125...130...136...

Well, these numbers are just numbers so interpreting them requires a protocol to set an agreement about what they mean. Whatever is receiving them needs to already know that it's getting brightness data, and that the range from 0 to 255 represents from dark to dazzling. There is no way of telling if there's an error in the data or determining which sensor might be sending them to us. But if we're just making a single lamp that matches the current brightness from a single sensor on the roof, this may be all we need. If we get a wrong number every once in a while, the worst that may happen is the lamp might flicker for a moment. No big deal! So in this case we're all set. But what if our roof sensor is getting both light and temperature data? How can we tell the difference?

Start bytes

If we're sending two pieces of data, the first solution you might come up with is to always send them in order, first light and then temperature, like this:

    136...14...137...14...137...14...138...14...138...15...138...15...139...15...

For clarity in this example we've made the temperature numbers much smaller than the light numbers, but we certainly couldn't count on that always being the case in real data. If we plugged in our lamp at some arbitrary moment, we might see the following instead:

    ...137...137...138...137...138...138...139...138...138...138...139...139...138...

Which number represents light and which number represents temperature? There isn't any way to be sure, and that's no good. We have to come up with a better solution. Maybe we could add a special number at the beginning of the sequence, like 255, so that every time we see it we'd know the very next number would be a light value, followed by a temperature value. That would look like this:

    255...136...14...255...137...14...255...137...14...255...138...14...255...138...15...

Great! Now our data is all organized in a sequence. The 255 in this case is known as a *start byte*. The start byte concept is so useful that you'll find it in many other protocols, including in the XBee's API. (By the way, we should make sure that our data values stop at 254 so that the start byte will always be unique.) For a computer to read this, we simply tell it to look for a 255, then read in the next byte as a light value and the third as a temperature value. It's a total solution, as long as the sequence and type of sensor values we're sending are fixed. But what if they aren't?

Length byte

What if sometimes our node sends light and temperature, but other times it sends light, temperature, and humidity? No problem. In this case we need to add a value to our protocol to indicate the length of data coming after the start byte:

```
255...2...136...14...255...3...137...14...87...255...3...137...14...89...
```

In the sequence above, the numbers 2 and 3 indicate the length of the data. So this is the *length byte*. Now when a computer reads the sequence, it can know without a doubt that after the start byte, it gets a number that tells it how many more data values to read in.

 Protocol structures like these are often described as *frames*, *packets*, or *envelopes*. Each of those terms means pretty much the same thing—a repeating sequence containing *useful data* (sometimes called the payload) packaged with information *about* the data (sometimes called metadata). We'll use the term frame from here on out to describe our hypothetical protocol.

Contents ID

The length byte is handy, but it doesn't fully ensure that we know what the frame of data contains. For example, maybe sometimes we have light, temperature, and humidity, but other times we're sending pressure, rainfall, and wind speed. In both cases there are three pieces of data, so we also need to describe the contents of each frame. The simple thing is to add a *contents byte*, a number that acts as an ID for the type of data in a particular frame. We can decide arbitrarily that 1 will indicate a light/temperature/humidity frame and that 2 will indicate a pressure/rainfall/windspeed frame:

```
255...3...1...137...14...87...255...3...2...119...28...54...
```

So our sequence here is start byte, length byte, ID byte, and then the data itself. This kind of predictable format is just what computers adore! It transmits everything we need to know, in as little space as possible. There's no limit to the type of useful meta-information we could add in this manner. For example, we might include an address byte to say which sensor node was sending the information, or a voltage byte to indicate the charge remaining in the sensor node's battery. As long as the sequence is predictable, it's an airtight method for communicating both the data and contextual information about that data.

There's one more item we should probably append to our message format. We've done a good job sequencing the numbers so that each one means something, but what if there's a transmission error? All methods of transmitting data are subject to corruption. Radio transmissions in particular are notoriously noisy. Static or interference of any kind could potentially introduce a stray bit into our data sequence. For example, in binary a single click of radio noise could easily turn a 21 into the number 149. While there isn't any way to prevent corruption like this from happening, there are thankfully many ways of detecting it. Error-correction schemes can be rather complex, but the concepts that they use are quite simple to understand.

Jane and Michael are having a conversation across a noisy room. Michael wants Jane to bring him a glass of wine, a napkin, and a celery stick. He could yell to her, "Bring me a glass of wine, a napkin, and a celery stick! Three things!" That last part is for error correction. If Jane only heard the glass of wine and the celery stick, that's not going to match up with "three things." In this case, she'd probably yell back, "What?" to let Michael know he needed to repeat himself. Computers use the same strategies that people do to detect problems in their communications. This particular method would be described as a *checksum*, meaning a *sum* of items used solely to *check* for communication errors. Computer protocols often use more sophisticated arithmetic than simply counting the items, but the principle of sending some frame information followed by a number that can be used to check the frame is widely employed to detect errors in everything from spacecraft communications to credit card numbers.

 It's important to note that checksums don't provide a guarantee of error-free communications. For example, if Jane brings Michael a beer, a fork, and a slice of cake, those are still three things, but definitely not the correct ones. More sophisticated checksums drastically decrease the probability of such "substitution" errors, but don't entirely eliminate them. For example, read about the cyclic redundancy check (CRC) at *http://wikipedia.org/wiki/Cyclic_redundancy_check*.

XBee API Protocol

Now that you know something about how protocols are designed, it should be fairly easy to understand the API format for XBee radios. The XBee API uses the same structures as our hypothetical protocol, and does so for the exact same reasons. The goal of API-mode communications is to transmit highly structured data quickly, predictably, and reliably. We will begin by taking a look at the structures shared by all API data frames and work our way into the specifics for each frame type.

The tiny microcontroller inside the Series 2 XBee radio doesn't have enough room to hold all the instructions for both AT mode (transparent/command modes) and API mode. Therefore, different firmware must be loaded onto the radio with X-CTU depending upon which mode you'd like to use to communicate over the local serial port. All the ZigBee firmware versions end in either AT or API (see Figure 5-2) to indicate how they will talk to you on their serial ports.

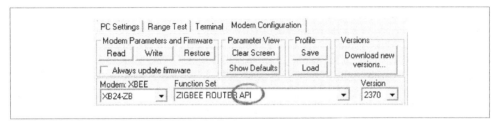

Figure 5-2. To use API mode, be sure to select a firmware function set that ends in "API"

The XBee API frame consists of a series of bytes, each new one building on the information already transmitted.

Let's dig in. You've already read about all the concepts that the API format uses, so hopefully each byte we discuss will now seem like an old friend. Table 5-1 shows the basic structure of the frame.

Table 5-1. Basic API frame structure

Start delimiter	Length		Frame data			Checksum
Byte 1	Byte 2	Byte 3	Byte 4	…	Byte n	Byte n+1
0x7E	MSB	LSB	API-specific structure			Single byte

Start Delimiter

Every API frame begins with a start byte. This is a unique number that indicates we are at the beginning of the data frame. In our hypothetical example above, we used decimal 255 for this. The XBee API employs decimal 126 for the exact same purpose. Because everything in the API documentation is described in hexadecimal format, we'll do that here, too. Remember that hex is just notation: decimal 126 and 0x7E are exactly the same number, just written down in different styles. (If you need to brush up on hexadecimals, this is a good time to flip back to the sidebar "Hexadecimals" on page 30 in Chapter 2.)

If we start reading bytes that are arriving from the XBee's serial port in midstream, we won't know what they represent until we know their order. So the first thing to do is look for a start byte of 0x7E. Once we get that, we know where we are and everything else can fall into place. The start byte is like the front cover of a book.

Length Bytes

The next two numbers we receive after the start byte indicate the overall length of the data frame. This lets us know how long to keep reading before we stop, in effect letting us know where the back cover of the book is. Right now the second byte, listed as MSB (*most significant byte*), is usually zero and the third one, listed as LSB (*least significant byte*), usually contains the entire length.

Because a very long data frame could exceed the number that can be described in a byte (remember, that's 0–255), we use a second byte to extend the value to a 16-bit number (0–65,535). In this case, the large part of the number will be covered in byte 2, the MSB, while the small part of the number would go into byte 3, the LSB. See the sidebar "Breaking Large Numbers into Bytes" on page 135 for more information.

Frame Data Bytes

The frame data is specific to each type of message we receive from the XBee radio. This is the guts of the information, and we'll expand on its internal structure below. Some frames will carry a great deal of internal data while the smallest frame contains only 2 bytes. For now, you can consider frame data to be like the inside pages of a book. Different kinds of books have different layouts, and the frame data functions in the same way. Remember that since we've read in the length byte, we already know exactly how long this frame data will be.

Checksum

The very last byte of the frame is always a checksum, so we can think of it as the back cover of our book. The checksum is calculated based on all the bytes that came before it. It's a simple sum of all the bytes that made up the frame, used at the receiving end to check and see if there was a transmission error. The calculation is regular arithmetic, designed to be extremely efficient for computers to process.

Here's the checksum formula, as stated in the official documentation:

- To calculate: Not including frame delimiters and length, add all bytes, keeping only the lowest 8 bits of the result, and subtract the result from 0xFF.
- To verify: Add all bytes (include checksum, but not the delimiter and length bytes). If the checksum is correct, the sum will equal 0xFF.

The checksum formula is mostly addition and subtraction, so it's very easy to program. (Keeping the lowest 8 bits of the result is accomplished in code with a *bit mask* operation that typically looks something like this: & 0xFF.) Usually you'd write a function in your program to do the whole checksum calculation for you. In most of our examples, even that isn't necessary because we use software libraries to do this work for us.

API Frame Types

Inside the general frame structure there are substructures that cover all the different kinds of data that you might want to send to and receive from your local XBee radio. Different types of frames contain different types of data structures in much the same way as different types of books contain different internal formats. When you pick up a cookbook, you expect to see a title page, an explanation of techniques, then a bunch of recipes (ingredients first), and finally a comprehensive index at the end. That's like one API frame type. A novel is totally different. After its title page, you expect to see a series of chapters, followed by an acknowledgments section that expresses gratitude to everyone ignored during the writing process, and finally a paragraph about the author's expensive schooling and trendy place of residence. The novel is like a second different API frame type. The cookbook and the novel both have front and back covers. Both have title pages. However, their internal structures follow different standardized patterns to help convey the different kinds of information the books contain.

There are more than a dozen different API frame types currently defined for the XBee ZB. We'll look at eight of them here.

The frame type byte tells us what type of API frame we are looking at. Knowing the frame type is crucial to knowing what information is coming next. For example, if the frame type is 0x08, that indicates it is an AT command frame. So by reading the first four bytes we will know:

- Where the frame begins (start byte)
- How long the frame is going to be (length bytes)
- What kind of frame we're looking at (frame type)

Every frame type is assigned a number. Table 5-2 lists the basic ones, including all the frame types we'll cover in this chapter.

Table 5-2. Some API mode frame types

Frame type	Description
0x08	AT command (immediate)
0x09	AT command (queued)
0x17	Remote Command Request
0x88	AT command response
0x8A	Modem Status

Frame type	Description
0x10	TX request
0x8B	TX response
0x90	RX received
0x92	RX I/O data received
0x95	Node Identification Indicator
0x97	Remote Command Response

AT Commands

AT-type commands can be sent via API frames to configure your local radio. They can query the settings on the local radio or set parameters. These are all the same commands you typed in transparent/command mode.

Just like all the other frame types, AT command frames begin with our old buddy the start byte: 0x7E (see Table 5-3). This is followed by two bytes that indicate the frame length. There's also a checksum at the end. The data that makes the AT command frame unique goes in the Frame-specific section, starting right after the length byte.

Table 5-3. API format for AT commands

Frame fields		Offset	Example	Description
Start delimiter		0	0x7E	
Length	MSB 1	0x00		Number of bytes between the length and the checksum.
	LSB 2	0x04		
Frame-specific data	Frame type	3	0x08	
	Frame ID	4	0x52	Identifies the UART data frame for the host to correlate with a subsequent ACK (acknowledgment). If set to 0, no response is sent.
	AT command	5	0x4E (N)	The command name—two ASCII characters that identify the AT command.
		6	0x4A (J)	
	Parameter value			If present, indicates the requested parameter value to set the given register.
	(optional)			If no characters present, the register is queried.
Checksum		7	0x0D	0xFF – the 8-bit sum of bytes from offset 3 to this byte.

Frame type

The AT command frame is identified with 0x08. This lets the receiving radio know that the bytes that follow are going to be in the AT command frame order.

Frame ID

Since we set a frame type of 0x08, the XBee receiving this data from us knows that the next byte contains a frame ID. The frame ID is simply a serial number that we attach to the command. Results will be tagged with the same ID. That way, if we've sent a number of commands, we can tell which ones came back OK and which ones might have been lost or gotten an error. By the way, if you set the frame ID to 0x0, you suppress any response from the XBee, but it will still carry out your command.

Generally, you'll set the frame ID to 0x1 for the first command you send, then 0x2 for the next one, and so forth, until you get to 0xFF (the largest number that a single byte can hold). At that point you can start over again with 0x1.

AT command

The two bytes that follow the frame ID contain the AT command itself. The letters AT are omitted. Since we already know this is an AT command frame, we just use the two-letter code of the command itself. For example, if we are sending the NJ command, the first byte will be the ASCII equivalent of a capital N in hexadecimal format: 0x4E. The second will be a capital J, notated as 0x4A.

 You can find the ASCII equivalents table for any character in the Appendix .

Parameter value

If the command you're sending requires a parameter, such as a specific register setting, those bytes will follow the AT command. If no parameter value is given, the command will be treated as a query, with the results sent back in a response frame as detailed below. Parameters that are larger than a single byte can be split across several bytes with the larger, "most significant" part of the split-up number coming first.

Checksum

The AT command frame, like all API frames, ends with a checksum as described above.

Here's what a whole AT command frame looks like. It's the same one as in Table 5-3:

```
0x7E...0x00...0x04...0x08...0x52...0x4E...0x4A...0x0D
```

Mostly you'll use the API format coded into a program you modify that talks directly over the serial port to the local XBee. Alternatively, you might work with a software library that is programmed to talk in API format. However, it's entirely possible to figure out the bytes yourself and type them manually into any terminal program that provides a hex-adecimal interface. Both X-CTU and CoolTerm have hex modes. In the X-CTU Terminal tab, click on Assemble Packet to be able to type in hex, and click on Show Hex to be able to see the responses from your radio formatted in hex. In CoolTerm, select Send String from the Connection menu and click on the Hex button to type in hex. Click on the View Hex button to see hex-formatted responses from the XBee.

AT Responses

In API mode, every AT command sent to a local XBee radio can receive a response back from the XBee that contains the status of the command and optionally the registry value if one was requested in a query. This is a frame that the radio generates so you will *read* these but will never write one yourself. Table 5-4 shows the response format.

Whether your program cares about these responses will depend on your particular context. In some cases, the quick-and-dirty method of simply sending commands and ignoring the responses is a perfectly serviceable solution. For example, if you are prototyping an error-tolerant interactive sculpture project, then dealing with AT command responses may be far more trouble than it's worth. On the off chance that an error happens, your audience might not even notice. Naturally there are other applications where you wouldn't want to be so tolerant. If you're leaving a sensor network out in the desert by itself for a year, every detail must be addressed with strict response processing and error handling. The important thing is to pick the level of thoroughness that's appropriate to your project and not go overboard without a good reason.

Table 5-4. API format for AT command responses

Frame fields		Offset	Example	Description
Start delimiter		0	0x7E	
Length		MSB 1	0x00	Number of bytes between the length and the checksum.
		LSB 2	0x05	
Frame-specific data	Frame type	3	0x88	
	Frame ID	4	0x01	Identifies the UART data frame being reported. Note: If frame ID = 0 in AT command mode, no AT command response will be given.

Frame fields		Offset	Example	Description
	AT command	5	'B' = 0x42	The command name—two ASCII characters that identify the AT command.
		6	'D' = 0x44	
	Command status	7	0x00	0 = OK
				1 = ERROR
				2 = Invalid Command
				3 = Invalid Parameter
				4 = Tx Failure
	Command data			Register data in binary format. If the register was set, this field is not returned, as in this example.
Checksum		8	0xF0	0xFF – the 8-bit sum of bytes from offset 3 to this byte.

AT command response frames received back from the local XBee should look fairly familiar by now. There's a start byte, length bytes, a frame type, and a frame ID, followed by the type of AT command you sent. This is followed by the command status and data, which we'll look at in detail. As you might expect, the last byte is a checksum, calculated in the usual way.

Frame type

The AT command response frame type is always 0x88.

Frame ID

The frame ID you get back will be the same as the one you sent with the original AT command request. You can use the ID to match up your request with this response. Remember that if you set your request frame ID to 0x0, you won't get any response frame in the first place.

AT command

These two bytes will be the ASCII equivalents of the two command characters you sent.

Command status

This next byte indicates how your command fared. 0x0 indicates that everything went fine. It's like receiving an OK in transparent/command mode and should cause you and your program to do a happy dance. A value of 0x1 indicates that your command resulted in an ERROR. This means it was recognized but could not be carried out for some reason. Receiving 0x2 indicates that your command itself was invalid. Maybe you got one of the letters wrong? A value of 0x3 indicates that the command was recognized but the parameters you sent with it were out of range. Finally, 0x4 indicates a transmission failure.

Command data

If you queried a register by sending a command with no parameters, these bytes will contain the response information. The response will be broken up into bytes and may represent a number or hex-encoded ASCII string.

 By the way, viewing a stream of API frames displayed as ASCII characters in a terminal program will look something like this:

```
~.......@R..].........F|~.......@R..].........F|~.......@R..
].........E}~.. .....@R..].........E}~.......@R..]........
F|~.......@R..].........F|~..... ..@R..].........E}~.......@R..
].........E}~.......@R..].........E}~.......@ R..]........
F|~.......@R..].........E}~.......@R..].........F|~.......@R..
].........E}~.......@R..].........E}~.......@R..].........F|
```

Note the repeating tilde (~) character. This is the ASCII equivalent of 0x7E, the start byte, and is a clear indication that rather than seeing garbage, you are seeing good data being delivered in API mode. Switch into viewing hex to see the API frame contents properly.

ZigBee Transmit Request

Let's send some real data! This frame is how you tell your local radio to send information to some other remote radio. The ZigBee Transmit Request frame encapsulates your payload information (the data itself) with a batch of addressing and transmission options that describe how the payload should be delivered. This frame is a great example of how API mode facilitates something that can't easily be accomplished in transparent/command mode: setting destination addresses on the fly. Now instead of issuing a +++ and a bunch of commands each time we want to change the destination address, we simply attach that destination to each a frame of data and send it on its way. This is a much more efficient process, especially if you have a network with hundreds of different nodes that you might need to use as destinations. Table 5-5 shows the ZigBee Transmit Request format.

Table 5-5. API format for ZigBee Transmit Request

Frame fields		Offset	Example	Description
Start delimiter		0	0x7E	
Length		MSB 1	0x00	Number of bytes between the length and the checksum.
		LSB 2	0x16	
Frame-specific data	Frame type	3	0x10	
	Frame ID	4	0x01	Identifies the UART data frame for the host to correlate with a subsequent ACK (acknowledgment). If set to 0, no response is sent.

Frame fields		Offset	Example	Description
		MSB 5	0x00	Set to the 64-bit address of the destination device. The following addresses are also supported: 0x0000000000000000 – Reserved 64-bit address for the coordinator. 0x000000000000FFFF – Broadcast address.
	64-bit destination address	6	0x13	
		7	0xA2	
		8	0x00	
		9	0x40	
		10	0x0A	
		11	0x01	
		LSB 12	0x27	
	16-bit destination network address	MSB 13	0xFF	Set to the 16-bit address of the destination device, if known. Set to 0xFFFE if the address is unknown, or if sending a broadcast.
		LSB 14	0xFE	
	Broadcast radius	15	0x00	Sets maximum number of hops a broadcast transmission can take. If set to 0, the broadcast radius will be set to the maximum hops value.
	Options	16	0x00	Bit field of supported transmission options. Supported values include: 0x01 – Disable ACK 0x20 – Enable APS encryption (if EE=1) 0x40 – Use the extended transmission timeout for this destination Enabling APS encryption decreases the maximum number of RF payload bytes by 4 (below the value reported by NP). Setting the extended timeout bit causes the stack to set the extended transmission timeout for the destination address. All unused and unsupported bits must be set to 0.
	RF data	17	0x54	Data that is sent to the destination device.
		18	0x78	
		19	0x44	
		20	0x61	
		21	0x74	
		22	0x61	
		23	0x30	
		24	0x41	
Checksum		25	0x13	0xFF – the 8-bit sum of bytes from offset 3 to this byte.

Again, our frame begins with a start byte, length bytes, a frame type (in this case 0x10, indicating the ZigBee Transmit Request format), and a frame ID. This preamble is followed by addressing information that we'll look at in detail, and then by the data payload itself. The frame concludes as always with a single-byte checksum.

64-bit destination address

These eight bytes indicate the unique-in-the-world destination address for this transmission, for example 0x0013A200400A0127. There are two special addresses that you can also use. If you want to reach the network coordinator, you can set this address to 0x0000000000000000 (that's 16 zeros) and it will be routed automatically. To send a broadcast message that is delivered to all nodes on the network, set the 64-bit destination address to 0x000000000000FFFF. Check Chapter 7 for information about the ATND node discovery command that can be used to discover all the 64-bit addresses currently present on the network.

16-bit destination network address

These two bytes can be set to the 16-bit address of the destination radio, if you know what that is. Assigning this address manually is optional, but it will greatly speed up your transmission. This can be *essential* on larger networks. See "Limits of 64-bit Addressing" on page 126 for a description of the lookup process. If you don't know the 16-bit address that the coordinator has assigned for the destination, simply set these two bytes to 0xFF and 0xFE respectively. This will cause an address lookup to occur so that the transmission can be properly delivered. 0xFFFE is also the proper 16-bit address setting for broadcast transmissions to be delivered to all the devices on the network.

Limits of 64-bit Addressing

Using 64-bit addressing to route messages requires broadcast transmissions to discover the 16-bit address. This is almost *never* a good idea when using the Series 2 on networks that are larger than around 10 nodes. Here's what happens during a transmission cycle when the 16-bit address is set to 0xFFFE for broadcast:

1. A broadcast is sent three times (a value controlled by the ZigBee stack profile) on the network asking to resolve the 64-bit address to the 16-bit network address. These broadcasts are very, very expensive in terms of routing and network overhead because they create three additional messages to every node on the network for every single message sent by any radio.

2. One or more nodes respond to the requester with a point-to-point frame containing the 16-bit address.

3. The transmission proceeds with the newly discovered 16-bit address being used.

If you have started with 64-bit addressing for your messages and your network grows, you will want to migrate your application toward either discovering and using the

---

16-bit addresses in advance via the API or saving them offboard on your computer or device when it receives incoming data from the remote node. (If you do this, also consider tracking the TX status of any transmissions using the short address to see if the transmission fails so that you can invalidate the known 16-bit address to 0xFFFE and start the process again.) Remember that you don't need to worry about any of this if your network is relatively small, if messages are not sent too frequently, or if you are using a ConnectPort X gateway—as this is handled for you automatically. Phew!

Broadcast radius. Set this to 0x0. Each broadcast message can be constrained to a certain radius, usually defined by the default broadcast timeout value set in ATNH. This is an advanced setting for dealing with very specific application or network issues. You should almost always leave this at 0x0 and use the defaults.

Options. Set this to 0x0. As of this writing, there are no options defined for this frame type, though future versions of the firmware might implement additional features using this byte.

RF data. At long last we come to the payload. The payload is the data we wanted to send in the first place! It is the meat of our protocol sandwich (or the tasty eggplant, in case meat isn't your thing). Assemble your data into a string of bytes. On many small networks you can usually put up to 84 bytes in your payload transmission. Of course, if you keep your individual data transmissions small, you won't need to worry about this limit.

> The exact number of allowed payload bytes in each frame is reduced when encryption or source routing are enabled (see Chapter 8). There's nothing in this book that requires you to use those features; however, if at some point in the future you decide to go with encryption or source routing, you can query the ATNP register to determine the current payload size limits for your network.

ZigBee Transmit Status

Another advantage to API mode is that transmissions don't just flow out into a virtual black hole. For each transmission where the frame ID is set to something other than 0x0, we receive back a full status report on any discovery, transmission, or delivery issues. Sometimes this doesn't matter one whit, especially if you're just doing a quick prototype or are running an application that's tolerant of an occasional failure. Transmissions to blink your holiday lighting don't require detailed status reports. Transmissions that monitor your home security probably do. Here's what that status message looks like. It contains all the now-familiar components (see Table 5-6). The frame type is set to 0x8B so you know it's a ZigBee Transmit Status. The frame ID will be the one you put in the original ZigBee Transmit Request that this Status frame is reporting on. There are also a few new components to indicate the transmit retry count, delivery status, and discovery status.

Table 5-6. API format for ZigBee Transmit Status

Frame fields		Offset	Example	Description
Start delimiter		0	0x7E	
Length	MSB 1	0x00	Number of bytes between the length and the checksum.
	LSB 2	0x07	
Frame-specific data	Frame type	3	0x8B	
	Frame ID	4	0x01	Identifies the UART data frame being reported. Note: If frame ID = 0 in AT command mode, no AT command response will be given.
	16-bit address destination	5	0x7D	If successful, this is the 16-bit network address the packet was delivered to. If not successful, this address matches the destination network address that was provided in the Transmit Request frame.
		6	0x84	
	Transmit retry count	7	0x00	The number of application transmission retries that took place.
	Delivery status	8	0x00	0x00 = Success 0x01 = MAC ACK failure 0x02 = CCA failure 0x15 = Invalid destination endpoint 0x21 = Network ACK failure 0x22 = Not joined to network 0x23 = Self-addressed 0x24 = Address not found 0x25 = Route not found 0x26 = Broadcast source failed to hear a neighbor relay the message 0x2B = Invalid binding table index 0x2C = Resource error, lack of free buffers, timers, etc. 0x2D = Attempted broadcast with APS transmission 0x2E = Attempted unicast with APS transmission, but EE=0 0x32 = Resource error, lack of free buffers, timers, etc. 0x74 = Data payload too large 0x75 = Indirect message unrequested
	Discovery status	9	0x01	0x00 = No discovery overhead 0x01 = Address discovery

Frame fields	Offset	Example	Description
			0x02 = Route discovery
			0x03 = Address and route
			0x40 = Extended timeout discovery
Checksum	10	0x71	0xFF – the 8-bit sum of bytes from offset 3 to this byte.

Transmit retry count

Every transmission will be attempted up to three times by the transmitting radio (other retries may happen invisibly along the mesh route). The count of these retries is listed in this byte. Retries are a normal part of wireless networking, so individually they are of no concern, though considered in aggregate they might indicate layout or interference issues. For now, there's no need to be particularly concerned about this count.

Delivery status

If this byte is 0x0, then hurray! Your transmission was successfully delivered to the destination address. Otherwise, the number you receive in this byte will indicate the kind of issue that prevented delivery, which is useful for debugging and possibly for deciding whether to send the information again. The error numbers are listed in Table 5-6. Many applications don't care why the error happened; they just need to know that it did. In this case, anything greater than 0x0 might tell your project to try the transmission again or to report an error to the user.

Discovery status

This byte gives a bit of information about how much overhead it took to discover the route for this transmission. In general, smaller numbers are better. For very large networks, you might want to keep an eye on this and consider using advanced source routing. For small networks like the ones we create in this book, the discovery status can be safely ignored.

ZigBee Receive Packet

Here's another API frame that gives us far more than we could get from simple transparent/command mode interactions. When a transmission is received in transparent mode, it comes with no indication of who the sender was. On a simple pair network that's fine because there's only one possible sender. But on a larger network, it's usually of considerable interest to know not only what was received but where it came from. So in addition to the usual preamble bytes, including the frame type of 0x90 to indicate a ZigBee Receive Packet, we get to see the 64-bit and 16-bit source addresses along with a receive options indicator, and of course the payload data itself, followed by a checksum. Table 5-7 shows this frame's format.

Table 5-7. API format for ZigBee RX Packet

Frame fields		Offset	Example	Description
Start delimiter		0	0x7E	
Length		MSB 1	0x00	Number of bytes between the length and the checksum.
		LSB 2	0x12	
Frame-specific data	Frame type	3	0x90	
	64-bit source address	MSB 4	0x00	64-bit address of sender. Set to 0xFFFFFFFFFFFFFFFF (unknown 64-bit address) if the sender's 64-bit address is unknown.
		5	0x13	
		6	0xA2	
		7	0x00	
		8	0x40	
		9	0x52	
		10	0x2B	
		LSB 11	0xAA	
	16-bit source network address	MSB 12	0x7D	16-bit address of sender.
		LSB 13	0x84	
	Receive options	14	0x01	0x01 – Packet acknowledged. 0x02 – Packet was a broadcast packet. 0x20 – Packet encrypted with APS encryption. 0x40 – Packet was sent from an end device (if known).
	Received data	15	0x52	Received RF data.
		16	0x78	
		17	0x44	
		18	0x61	
		19	0x74	
		20	0x61	
Checksum		21	0x0D	0xFF – the 8-bit sum of bytes from offset 3 to this byte.

64-bit source address

These eight bytes report the address that this transmission was sent from. It's how we can tell which radio is associated with the data we just received.

16-bit source network address

These two bytes tell us the short network address of the sender. Feel free to ignore this for now, but keep in mind that later it could be handy in case we want to speed up the reply process. If the 16-bit address is included in a future transmission frame, we can save time and some overhead by not forcing the network to look it up again.

Receive options

This byte provides just a little info. It indicates 0x1 if receipt of transmission was acknowledged, or 0x2 if the received information was sent as a broadcast, in which case no acknowledgment will be sent. In most cases this byte can be safely ignored.

Received data

This is the data itself, organized as bytes in the exact same order it was in when the sender sent it. This data, by the way, could be anything from a doorbell push indicator to a poem. We refer to it as arbitrary data. It isn't arbitrary in the sense that it is random but rather because it doesn't need to follow a specific structure.

I/O Data Sample Rx Indicator

We've covered all the basic API frame types used to issue local commands, transmit information, receive information, and check on the status of our commands and transmissions. Since you've made it this far, it should be pretty easy to understand the next API frame type. It contains the juiciest type of information—direct sensor data! This is how you will obtain real values from networks of remote sensors via the XBee's direct input/output functionality. Your room temperature, soil moisture, monkey-trap status, or whatever, will arrive encased in this frame type.

The ZigBee I/O Sample Rx Indicator is really just an extension of the ZigBee Receive Packet discussed above. The main difference is that instead of the payload having an arbitrary or unconstrained format, it is organized in a highly structured way that lets us decode a set of digital and/or analog samples that were taken directly by the transmitting XBee. It's important to note that I/O samples can't be received in transparent/command mode at all. Using API mode is essential to receiving XBee direct I/O information. It's one of the most important reasons for us to cover the API in this book.

All I/O samples are received inside what otherwise would appear to be a simple ZigBee Receive Packet. The first clue that it's any different is its frame type of 0x92, which indicates that we'll be getting an I/O data sample in the payload. After that, everything is the same up until the first payload byte, which gives us the number of samples followed by the analog and digital channel masks that tell us how the sender's pins are configured. Then the digital and analog samples themselves are provided, all in a highly structured format that allows the data, when correctly interpreted, to be absolutely unambiguous. Table 5-8 shows the format for this frame.

Table 5-8. API format for ZigBee I/O Data Sample Rx Indicator

Frame fields		Offset	Example	Description
Start delimiter		0	0x7E	
Length		MSB 1	0x00	Number of bytes between the length and the checksum.
		LSB 2	0x14	
Frame-specific data	Frame type	3	0x92	
		MSB 4	0x00	
		5	0x13	
		6	0xA2	
	64-bit source address	7	0x00	64-bit address of sender.
		8	0x40	
		9	0x52	
		10	0x2B	
		LSB 11	0xAA	
	16-bit source network address	MSB 12	0x7D	16-bit address of sender.
		LSB 13	0x84	
	Receive options	14	0x01	0x01 – Packet acknowledged.
				0x02 – Packet was a broadcast packet.
	Number of samples	15	0x01	Number of sample sets included in the payload. (Always set to 1.)
	Digital channel mask	16	0x00	Bit mask field that indicates which digital I/O lines on the remote have sampling enabled (if any).
		17	0x1C	
	Analog channel mask	18	0x02	Bit mask field that indicates which analog I/O lines on the remote have sampling enabled (if any).
	Digital samples (if included)	19	0x00	If the sample set includes any digital I/O lines (digital channel mask > 0), these two bytes contain samples for all enabled digital I/O lines.
		20	0x14	
				DIO lines that do not have sampling enabled return 0. Bits in these two bytes map the same as they do in the Digital Channels Mask field.
	Analog sample	21	0x02	If the sample set includes any analog input lines (analog channel mask > 0), each enabled analog input returns a 2-byte value indicating the A/D measurement of that input. Analog samples are ordered sequentially from AD0/DIO0 to AD3/DIO3, to the supply voltage.
		22	0x25	
Checksum		23	0xF5	0xFF – the 8-bit sum of bytes from offset 3 to this byte.

Number of samples

This single byte indicates how many sampling collections are contained in this frame. Currently this is always set to 0x1 to indicate a single collection, because multiple collections are not yet supported on the Series 2 hardware. You can safely ignore this byte.

Digital channel mask

These two bytes indicate which of the sending XBee's pins are configured as digital inputs. Each hexadecimal can be translated to a binary number that will tell you which pins are configured as digital inputs. See the sidebar "Mapping Binary to Switches and Pins" on page 134 for information about how to do that. Once the numbers have been translated, you can read them using Tables 5-9 and 5-10.

Table 5-9. First byte of digital channel mask

n/a	n/a	n/a	D12	D11	D10	n/a	n/a

Table 5-10. Second byte of digital channel mask

D7	D6	D5	D4	D3	D2	D1	D0

As an example, let's say you received 0x0 as the first digital channel mask byte and 0x1C as the second one. The first byte in binary is 0000 0000. (The space in the middle doesn't mean anything; it just makes the number easier to look at.) Using Table 5-11, we can see that none of pins D12, D11, and D10 are configured as digital inputs because they're all set to zero. 0 means that pin is not configured as a digital input; 1 means that it is. The second byte, 0x1C, translates to 0001 1100 in binary. Placing this number into Table 5-12, we can see that pins D4, D3, and D2 are configured to be digital inputs because that's where the 1s show up.

Table 5-11. Example: first byte of digital channel mask showing that pins D10–D12 are NOT configured as digital inputs

n/a	n/a	n/a	D12	D11	D10	n/a	n/a
0	0	0	0	0	0	0	0

Table 5-12. Example: second byte of digital channel mask showing that ONLY pins D2–D4 are configured to be digital inputs

D7	D6	D5	D4	D3	D2	D1	D0
0	0	0	1	1	1	0	0

Mapping Binary to Switches and Pins

The XBee API uses an elegant, if somewhat commonplace, trick to accomplish representing pin states as hexadecimals—using the arrangement of ones and zeros in binary notation to directly describe pins as being either on or off. We discussed binary briefly earlier in this chapter as a method for indicating a certain number of states. We can also map a number directly to a set of switches (or pins in this case) that can be either on or off. So if you had eight switches that were all off you could represent their on/off state with 0000 0000. (Remember, the space is just for readability.) If the first, third, and eighth switches were flipped on, the mapped number would look like 1000 0101. (Binary numbers increase from right to left, just like decimal numbers.) Translating that map from a binary number into decimal notation gives us 133, which in hexadecimal notation is 0x85. So by receiving the number 0x85 and looking at its binary equivalent, we can know which switches are on and off. Another example: if we received the hex number 0x1C, we could translate that to binary notation of 0001 1100. So that means the third, fourth, and fifth switches are on. Remember that there are calculators available on most computers that will do the hex-to-binary translation for you.

Analog channel mask

There is only one byte for the analog channel mask. This is because we have only four analog inputs to consider. This mask uses the same system as the digital ones (see Table 5-13).

Table 5-13. Single byte for analog channel mask

(voltage)	n/a	n/a	n/a	A3	A2	A1	A0

Using Table 5-13, we can decode the binary version of the byte into a pin configuration. For example, if we received 0x2 as the analog channel mask, that would translate into the binary number 0000 0010. This would indicate that pin A1 is configured as an analog input, but none of the other pins are.

You may note that the highest bit in the analog channel mask indicates if system voltage readings have been enabled as part of the analog data set. By default they are not.

Digital samples

If you are receiving digital samples, these two bytes will appear and let you know whether the enabled pins are currently high or low, in the same arrangement as the mask. So if you receive 0x0 in the first digital sample byte and 0x14 in the second one, that indicates high voltage is being received only on pins D4 and D2. Any other pins configured as digital inputs are currently reading low. You can use the preceding tables to decode these samples, just like with the mask.

These two bytes will be received *only* if at least one pin is enabled as a digital input. If your digital channel mask bytes are both 0x0, no pins have been enabled and these two bytes will be *omitted*.

Analog samples

The last component of the I/O Data Sample frame is a set of two bytes for each analog sample taken. We know how many to expect from the analog channel mask, which tells us what pins have been configured as analog inputs. So if data is being received from two analog pins, we can expect to receive four bytes of data in all. Each two-byte sample consists of a most significant and a least significant byte. This is because the sample itself is represented as a 10-bit number. 10 bits are enough to represent values from 0–1,023, which gives us a pretty smooth resolution for our data.

Breaking Large Numbers into Bytes

Large numbers that are broken up into bytes for transmission can easily be reassembled once they are received. For example, the number 987 can't fit in a single byte. When the XBee radio needs to transmit this 10-bit number, it breaks it into two bytes. The lower part represents the part of the binary number that falls into the place values 0–255. The higher part represents the place values from 256 to 1,023. We receive 0x3 as the first, most significant byte (MSB) and 0xDB as the second, least significant byte (LSB). Pasting these together gives us 0x3DB, and the decimal equivalent of that is 987. In code, we can accomplish this paste process arithmetically: multiplying the MSB 0xFF (same as decimal 255) and then adding the result to the LSB will give us the correct results:

```
( 0x3 * 0x100 ) + 0xDB = 0x3DB ...same as 987 in decimal
```

Remote AT Command Request

Sending commands to configure the local radio is useful. Sending commands over the wireless network to configure remote radios is kind of exhilarating. It is also something you can accomplish only in API mode—yet another reason to master the API.

Any AT-type command that you can issue locally can also be sent wirelessly for execution on a remote radio. This is especially useful for remote actuation, where you might want to change the state of a digital output from low to high to trigger a real-world action. We'll use this command type to do just that in the next chapter. Table 5-14 shows the format of the Remote AT Command Request.

Table 5-14. API format for Remote AT Command Request

Frame fields		Offset	Example	Description
Start delimiter		0	0x7E	
Length		MSB 1	0x00	Number of bytes between the length and the checksum.
		LSB 2	0x10	
Frame-specific data	Frame type	3	0x17	
	Frame ID	4	0x01	Identifies the UART data frame for the host to correlate with a subsequent ACK (acknowledgment). If set to 0, no response is sent.
	64-bit destination address	MSB 5	0x00	Set to the 64-bit address of the destination device. The following addresses are also supported:
		6	0x13	
		7	0xA2	0x0000000000000000 – Reserved 64-bit address for the coordinator
		8	0x00	
		9	0x40	0x000000000000FFFF – Broadcast address
		10	0x40	
		11	0x11	
		LSB 12	0x22	
	16-bit destination network address	MSB 13	0xFF	Set to the 16-bit address of the destination device, if known. Set to 0xFFFE if the address is unknown, or if sending a broadcast.
		LSB 14	0xFE	
	Remote command options	15	0x02 (apply changes)	Bit field to enable various remote command options. Supported values include: 0x01 – Disable ACK. 0x02 – Apply changes on remote. (If not set, AC command must be sent before changes will take effect.) 0x40 – Use the extended transmission timeout for this destination. Setting the extended timeout bit causes the stack to set the extended transmission timeout for the destination address. All unused and unsupported bits must be set to 0.
	AT command	16	0x42 (B)	The name of the command.
		17	0x48 (H)	
	Command parameter	18	0x01	If present, indicates the requested parameter value to set the given register. If no characters present, the register is queried.
Checksum		19	0xF5	0xFF – the 8-bit sum of bytes from offset 3 to this byte.

The Remote AT Command Request is made up almost entirely of components we have already covered. The frame type is 0x17, followed by an ID, then the 64-bit and 16-bit addresses. The next byte is for remote command options, which we will look at below. That's followed by two bytes for the two characters of the command, one or more bytes to contain any parameter being sent, and finally the checksum. This is starting to get easy!

Remote command options

This byte can currently be set to one of two states. Normally you should set it to 0x02 to indicate that any changes requested by this remote AT command should be applied immediately. Occasionally you might *not* want to apply a command until a specific moment, for example if you wanted a bunch of output pins to all change at the same instant. To delay command execution, set this byte to be 0x0 and then issue an AC command later when you are ready for all your changes to take effect.

 Applying a set of changes all at once is known in the computing world as *atomicity*. As with the original conception of an atom as an indivisible bit of matter, atomicity is used when each change can't be separated out from the others.

Remote Command Response

Every Remote Command Request that is sent with a frame ID other than zero will receive a response frame to report on how the remote command fared. This frame type is 0x97, followed by the ID of the request, 64- and 16-bit address, the AT command you sent, a command status (just like the one for local AT commands), command data if you queried a register, and finally the checksum. There is nothing at all in this frame that is new to you. In fact, you may start to consider yourself something of an API expert! Table 5-15 shows the Remote Command Response format.

Table 5-15. API format for Remote Command Response

Frame Fields		Offset	Example	Description
Start delimiter		0	0x7E	
Length		MSB 1	0x00	Number of bytes between the length and the checksum.
		LSB 2	0x13	
Frame-specific data	Frame type	3	0x97	
	Frame ID	4	0x55	This is the same value passed in to the request.
	64-bit source (remote) address	MSB 5	0x00	The address of the remote radio returning this response.
		6	0x13	
		7	0xA2	
		8	0x00	

Frame Fields		Offset	Example	Description
		9	0x40	
		10	0x52	
		11	0x2B	
		LSB 12	0xAA	
	16-bit source (remote) address	MSB 13	0x7D	Set to the 16-bit network address of the remote.
		LSB 14	0x84	Set to 0xFFFE if unknown.
	AT commands	15	0x53	The name of the command.
		16	0x4C	
	Command status	17	0x00	0 = OK
				1 = ERROR
				2 = Invalid Command
				3 = Invalid Parameter
				4 = Remote Command Transmission Failed
	Command data	18	0x40	
		19	0x52	Register data in binary format. If the register was set, then this field is not returned.
		20	0x2B	
		21	0xAA	
Checksum		22	0xF0	0xFF – the 8-bit sum of bytes from offset 3 to this byte.

Using What You Need

Now that you understand how API mode works and how the structures around it function, we can talk about using it. It's pretty simple to write your own microcontroller code to work with the API, especially if you use only what you need. While the API is capable of supplying your application with an airtight protocol that covers any possible radio configuration, in most cases you'll be using only a small subset of that radio's capabilities. For example, in many sensor network applications your I/O data samples will all be the same length. This means you can get away with never checking the length byte. It also means that your data will show up in the same place in every frame. There's no need to calculate the digital mask or look for digital samples if you already know that your network never uses digital inputs.

The romantic lighting sensor example in the previous chapter takes advantage of this minimalist strategy. Here's the code from the loop that reads the I/O Data Sample frame:

```
// make sure everything we need is in the buffer
  if (Serial.available() >= 21) {
    // look for the start byte
```

```
    if (Serial.read() == 0x7E) {
      // blink debug LED to indicate when data is received
      digitalWrite(debugLED, HIGH);
      delay(10);
      digitalWrite(debugLED, LOW);
      // read the variables that we're not using out of the buffer
      for (int i = 0; i<18; i++) {
        byte discard = Serial.read();
      }
      int analogHigh = Serial.read();
      int analogLow = Serial.read();
      analogValue = analogLow + (analogHigh * 256);
    }
  }
```

We start by seeing if there's potentially a full frame of information waiting in the buffer. Because we set the sensor radio's configuration ourselves, we already know how long a frame will be: 22 bytes, or in this case 21 because we ignore the checksum. So we begin by checking the Arduino's serial buffer to see if there's potentially a full frame of data waiting for us:

```
// make sure everything we need is in the buffer
  if (Serial.available() >= 21) {
```

and then we read a byte in to see if it is a start byte of 0x7E. If it isn't we'll skip to the next byte until we do find a 0x7E and know we're most likely at the beginning of a data frame:

```
// look for the start byte
if (Serial.read() == 0x7E) {
```

When we know we're at the start of a data frame, we can skip over most of the frame contents. Length we already know. Same goes for frame type; we're only sending I/O sample frames so we can assume that whatever we receive will be the reply to one. The source address will always be the paired sensor, so that information can be ignored as well in this case. The receive options, number of samples, and all the channel mask information will never change in this project so we merrily read in all those bytes and throw them away:

```
// read the variables that we're not using out of the buffer
      for (int i = 0; i<18; i++) {
        byte discard = Serial.read();
      }
```

Finally we get to the bytes we do care about, the analog sensor readings from the remote photocell. The two bytes (MSB and LSB) get read into two variables:

```
      int analogHigh = Serial.read();
      int analogLow = Serial.read();
```

...and then pasted together arithmetically to reconstitute the original 10-bit sensor reading:

```
      analogValue = analogLow + (analogHigh * 256);
```

This whole process is very fast to program and will run just fine. Granted, there are more careful ways to go about this. A commercial application might want to do some additional error checking. In this case, however, we simply write the code that suits our romantic lighting purposes and get on with making the evening's fondue.

You may also want to take a close look at the setRemoteState function in "Program the romantic lighting sensor with feedback base station" on page 106. That function issues a Remote AT Command Request to turn on an LED that is connected to the remote sensor XBee in that project. It has all the elements that you are now familiar with, so the code comments probably make a great deal of sense to you now. Here is that function again. Read it through and admire just how technically adept you have become:

```
void setRemoteState(int value) {  // pass either a 0x4 or 0x5 to turn
                                  // the pin on or off
  Serial.print(0x7E, BYTE); // start byte
  Serial.print(0x0, BYTE);  // high part of length (always zero)
  Serial.print(0x10, BYTE); // low part of length (the number of bytes
                            // that follow, not including checksum)
  Serial.print(0x17, BYTE); // 0x17 is a remote AT command
  Serial.print(0x0, BYTE);  // frame id set to zero for no reply

  // ID of recipient, or use 0xFFFF for broadcast
  Serial.print(00, BYTE);
  Serial.print(00, BYTE);
  Serial.print(00, BYTE);
  Serial.print(00, BYTE);
  Serial.print(00, BYTE);
  Serial.print(00, BYTE);
  Serial.print(0xFF, BYTE); // 0xFF for broadcast
  Serial.print(0xFF, BYTE); // 0xFF for broadcast

  // 16 bit of recipient or 0xFFFE if unknown
  Serial.print(0xFF, BYTE);
  Serial.print(0xFE, BYTE);
  Serial.print(0x02, BYTE); // 0x02 to apply changes immediately on remote

  // command name in ASCII characters
  Serial.print('D', BYTE);
  Serial.print('1', BYTE);

  // command data in as many bytes as needed
  Serial.print(value, BYTE);

  // checksum is all bytes after length bytes
  long sum = 0x17 + 0xFF + 0xFF + 0xFF + 0xFE + 0x02 + 'D' + '1' + value;
  Serial.print( 0xFF - ( sum & 0xFF) , BYTE ); // calculate the proper checksum
}
```

Libraries

Another way to take advantage of the API is to use a software library that parses the API frames for you and presents the internal information in a slightly more human-friendly format. The upside is that everything is already written for you. The downside is that you may need to dig through considerable documentation to find the methods and attributes that apply to your situation. And you will definitely be glad that you know a thing or two about API frame format because most of these libraries use structures and terminology that exactly mirror it:

Arduino & C/C++

> This library by Andrew Rapp offers full support for both Series 1 and Series 2 XBee hardware in Arduino. It can be ported relatively easily into pure C/C++ environments as long as they support serial available/read/write/flush. (*http://code.google .com/p/xbee-arduino/*)

Processing & Java

> Another good library by Andrew Rapp is written for use in Java and can be ported over to work well in Processing, as we have done for the example that concludes this chapter. (*http://code.google.com/p/xbee-api/*)

Python

> Amit Snyderman created a library for Python environments that has been developed with significant contributions from several other developers. It requires the pySerial library (*http://pyserial.sourceforge.net/*). (*http://code.google.com/p/python -xbee/*)

Max/MSP

> A community-developed patch is available for reading API info into the Max/MSP graphical programming environment for multimedia. (*http://www.faludi.com/ xbee/max*)

PureData

> A similar patch has been ported over to the open source Pure Data multimedia graphical programming environment. (*http://www.faludi.com/xbee/pd*)

 You should be aware that there are two slightly different API frame specifications used by these libraries. The first one is the default, preselected with ATAP 1. The second can be selected by setting ATAP to 2, and it uses what are called "escaped" characters that avoid any possible confusion between data and control characters. The Java library we use in the example below does employ API operation with escaped characters. However, outside of being sure to select the proper ATAP setting when you configure your coordinator radio, you won't need to do anything differently in your code since the library will handle all the escaping for you.

Simple Sensor Network

This project can serve as a model for almost any sensor network you'd like to build. You will create a set of inexpensive temperature sensors that are mesh-networked together to stream their data to a base station radio. This base station will be connected to a computer where the real-time temperature data will be visualized on the screen. In the next chapter we will discuss making the sensor nodes very power-efficient so that they can be run effectively from batteries. For now, these nodes can be powered from wall outlets so we can concentrate on the business of building our first complete wireless sensor network.

> The example project in Figure 5-3 shows two sensor nodes and a base station. That's three radios in total. If you have only two radios, you can build it with a single sensor node and the base station. You can also create many more sensor nodes. If your network has more than 10 nodes, remember to extend the Processing code to change the display size and limits so that it can show all the data.

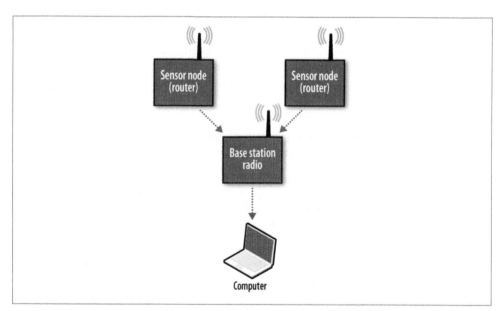

Figure 5-3. Simple sensor network

Parts

- Two solderless breadboards (AF 64, DK 438-1045-ND, SFE PRT-09567)
- Hookup wire or jumper wire kit (AF 153, DK 923351-ND, SFE PRT-00124)

- Two 9-volt or 5-volt power supplies (9-volt batteries also work well for short-term use) (AF 63, or 80 with 9 V battery, RS 273-355, SFE TOL-08269 or TOL-00298)
- Two 3.3 V voltage regulators (TO-220 package) (DK 497-1491-5-ND, SFE COM-00526)
- Two DC power jacks (2.1 mm ID, 5.5 mm OD) (DK CP-024A-ND, RS 274-1577, SFE PRT-00119)
- Two 100K ohm resistors (DK P100KBACT-ND, RS 271-1347, SFE has an assorted resistor kit: COM-09258)
- Two 200K ohm resistors (DK P200KBACT-ND, or use two 100K resistors in series for each board)
- Two 300 ohm resistors (DK P300BACT-ND, RS 271-012)
- Two 10 µF electrolytic capacitors (DK P966-ND, RS 272-1025, SFE COM-00523)
- Two 1 µF electrolytic capacitors (DK P993-ND, RS 272-1434)
- Two LM335 temperature sensors (in TO-92 packaging) (DK 497-2977-5-ND, SFE SEN-09438)
- One XBee radio (Series 2/ZB firmware) configured as a ZigBee Coordinator API mode (Digi: XB24-Z7WIT-004, DK 602-1098-ND)
- Two XBee radios (Series 2/ZB firmware) configured as a ZigBee Router AT mode (Digi: XB24-Z7WIT-004, DK 602-1098-ND)
- Two XBee breakout boards with male headers and 2 mm female headers installed (AF 126 (add SFE PRT-00116), SFE BOB-08276, PRT-08272, and PRT-00116)
- XBee USB serial adapter (XBee Explorer, Digi Evaluation board, or similar) (AF 247, SFE WRL-08687)
- USB cable for XBee adapter (AF 260, SFE CAB-00598)
- Wire strippers (AF 147, DK PAL70057-ND, SFE TOL-08696)

Prepare Your Coordinator Radio

1. Follow the instructions under "Reading Current Firmware and Configuration" on page 35 in Chapter 2 to configure one of your radios as a ZigBee Coordinator API. Note that your *coordinator* radio *must* use the API firmware for this project to work because I/O data is only delivered in API mode. Be sure to select the API version for your coordinator!

 When you change from AT to API mode using X-CTU, you may get an error message that the radio is no longer communicating. Go back to the PC Settings tab and check the Enable API box to enable communications with your radio. When you later change API mode to 2, go back to that tab and choose "Use escape characters (ATAP = 2)."

2. Once a radio has been set to API mode it can *only* be configured in X-CTU. You will not be able to make adjustments to this radio's configuration in CoolTerm. Use X-CTU to configure the coordinator with a PAN ID (between 0x0 and 0xFFFFFFFFFFFFFFFF) that you've selected, then click Write. Write down this PAN ID so you can program your router radio with the same one. Every radio in your network must use the same PAN ID so that they can communicate with each other (there's no need to set the DH and DL in this case, because the coordinator will only be receiving data, not sending it):

Pan ID:

_____

3. The software libraries that we are using in Processing *require* that the base station XBee be in API Mode 2 (API Operation with escaped characters). Use X-CTU to set AP (API Enable) to 2, and Write the configuration to your radio.

 Be sure that you set the coordinator's API to mode 2, otherwise the project will not work!

Prepare Your Router Radios

1. Follow the instructions under "Reading Current Firmware and Configuration" on page 35 in Chapter 2 to configure each of your sensor node radios as a ZigBee Router AT. Your *router* radios will use the *AT* firmware so that you can easily configure them using a serial terminal. Be sure you select the AT version for your routers!

 When you change from an API radio to an AT radio, you may get an error message that the radio is no longer communicating. If so, go back to the PC Settings tab and *un*check the Enable API Mode box.

2. Label the coordinator radio with a "C" so that you know which one it is later on. Each router radio can be labeled with an "R."

Prepare the Sensor Boards

We'll begin by configuring the router XBees. We'll use the CoolTerm terminal program and an XBee Explorer USB adapter again to set up your radios. For each of your sensor node radios:

1. Select a router XBee you've labeled with an "R" and place it into the XBee Explorer.

2. Plug the XBee Explorer into your computer.

3. Run the CoolTerm program and press the Options button to configure it.

4. Select the appropriate serial port, and check the Local Echo box so you can see your commands as you type them.

5. Click on the Connect button to connect to the serial port.

6. Type **+++** to go into command mode. You should receive an OK reply from the radio.

7. Select the *same* PAN ID you entered for your first radio above.

8. Type **ATID** followed by the PAN ID you selected and press Enter on the keyboard. You should receive OK again as a reply.

9. Every ZigBee coordinator always has 0 as its 16-bit network address. In fact, that's the default destination address for any newly configured XBee radio. To use 16-bit addressing, the high part of your radio's *destination* address will be zero. Type **ATDH 0** and press Enter on the keyboard. You should receive an OK response.

10. Enter **ATDL** followed by the *low* part of your radio's *destination* address, in this case also a zero because that's the fixed address for the coordinator. Type **ATDL 0** and press Enter. You should receive an OK response.

11. Enter **ATJV1** to ensure that your router attempts to rejoin the coordinator on startup.

12. Enter **ATD02** to put pin 0 in analog mode.

13. Enter **ATIR3E8** to set the sample rate to 1,000 milliseconds (hex 3E8).

14. Save your new settings as the radio's default by typing **ATWR** and pressing Enter.

 It's not a bad idea to recheck your configurations after you enter them. For example, to recheck that you entered the destination address correctly, from command mode type **ATDL** and press Enter to see the current setting.

Connect voltage regulator circuit and power jack to breadboard

1. Wire up a breadboard with a 3.3-volt voltage regulator (LD1117V33) as shown in Figure 5-4. The regulator has three legs—typically, ground, output, and input—when viewed from the front (where the writing is). Sometimes these legs are in a different order, so find and check the data sheet if you're not sure! Input is where a high voltage, for example 5 or 9 volts, is applied to the regulator. Output is where you will get the regulated 3.3 volts. Ground is the common ground for your entire circuit, including input, output, and all the other components. Bring ground out to both blue ground rails that run along the sides of your breadboard. Bring 3.3-volt output power to both of the red power rails.

Figure 5-4. Voltage regulator circuit on breadboard

2. Solder a red wire (about 10 cm) to the short center pin of your power jack, and solder a similar black wire to the longer outer pin, as shown in Figure 5-5. Don't allow the two connections to touch each other since that will create a short circuit when you power up!

3. Attach the red wire from the power jack, using the breadboard to connect it to the *input* pin of the voltage regulator. Attach the black ground wire to the ground pin of the voltage regulator in the same way.

4. Hook up the *output* pin of the voltage regulator to one of the power rails of the breadboard using a red wire. Hook up the ground pin to one of the ground rails on the breadboard.

5. Use the two capacitors to "decouple" the power supply in the following way: attach the short ground lead of the 10 μF capacitor (also marked with a stripe on the capacitor's ground side) to ground near the voltage regulator. Attach the other positive lead of the 10 μF capacitor to the voltage regulator's *input* pin. This will remove some lower-frequency noise coming from the wall power supply. Also attach the short ground lead of the 1 μF capacitor to ground, and the other positive lead to the 3.3 V *output* pin. This will remove some higher-frequency noise coming out of the voltage regulator. Decoupling will prevent noisy power from reaching your radio and interfering with its signal.

Figure 5-5. Power jack with wiring soldered in place

6. Hook up power and ground across the breadboard so that the rails on both sides are live.

 It's a really good idea to check the voltage levels using a multimeter after you first wire up the breadboard for power. Make sure that your power rails have 3.3 volts on both sides where you expect it. You don't want to send 9 volts to your radio and cook it!

Router XBee connection to power

1. With a *router* XBee mounted on its breakout board, position the breakout board in the center of your other breadboard so that the two rows of male header pins are inserted on opposite sides of the center trough.

2. Use red hookup wire to connect pin 1 (VCC) of the XBee to 3.3-volt regulated power.

3. Use black hookup wire to connect pin 10 (GND) of the XBee to ground.

Temperature input

This project uses the LM335 precision analog temperature sensor. This sensor has a linear output of +10 mV per degree Kelvin. It has an adjustment pin for calibration, but this can be safely ignored unless you mind one or two degrees of error at room temperature. You can get the entire data sheet for the LM335 at *http://www.national .com/ds/LM/LM135.pdf*:

1. The LM335 temperature sensor has three leads. When the sensor's flat side is facing you, the leads from left to right are adjustment, positive, and negative. Insert the LM335 so that each lead is in its own row on the breadboard.

2. Use a black wire to connect the rightmost, negative lead to one of the ground rails.

3. Insert the 300 ohm resistor so that it's connected to power on one end and to the positive center pin of the LM335 on the other. You can use jumper wires to make the connection if that's more convenient in your breadboard layout.

4. Insert a 200K ohm resistor so that it's connected to the positive center pin on one end and to an empty breadboard row on the other.

5. Use a jumper wire to connect between the unattached end of the 200K ohm resistor to XBee digital input 0 (physical pin 20).

6. Insert a 100K ohm resistor so that one end connects to XBee digital input 0 (and the 200K resistor). The other end of the 100K ohm resistor goes to ground. Again, use jumper wires as needed to complete these electrical connections.

Second sensor board

Create the second sensor board in the same way as the first. You can make as many sensor boards as you like. The system will work with as few as 1 or as many as 10 without any adjustment to the software. Figure 5-6 shows the breadboard layout for our simple sensor network, and Figure 5-7 shows the schematic.

Prepare the Base Station

Connect to computer

Your base station radio is simply an XBee serial adapter connected to your computer:

1. Select the coordinator XBee you've labeled with a "C" and place it into the XBee Explorer.

2. Plug the XBee Explorer into your computer.

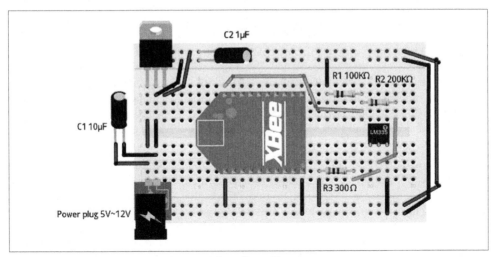

Figure 5-6. Simple sensor network LM335 breadboard layout

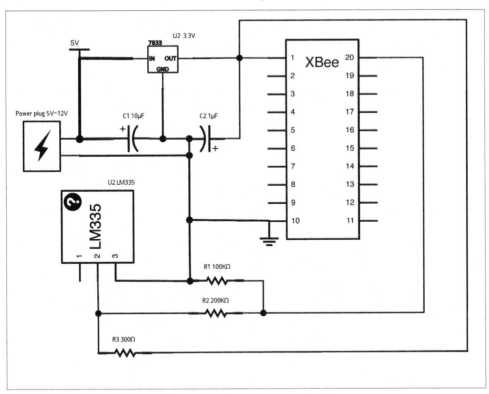

Figure 5-7. Simple sensor network LM335 schematic

All About Processing

Processing is an open source software development environment designed for novice programmers and geared toward visual displays. It was originally created as a way of teaching beginning programmers and has evolved into a highly popular environment for artists, interaction designers, software hackers, students, and professionals to create visually oriented applications. Processing is available for free, under the GNU General Pubic License. It operates on Macintosh, Windows, and Linux and can export both fully functional web applets and standalone applications for all three platforms.

You can download Processing from *http://processing.org/download* (see Figure 5-8). On the Mac, you'll get a disk image (*.dmg*) file that contains the Processing application. Simply drag it to your local *Applications* folder to install it and then double-click on the Processing icon to start the program. Windows downloads come in the form of a *.zip* file. Double-click the *.zip* file to open it, then drag the Processing folder into the *Program Files* directory (or any other location on your hard drive) and double-click *processing.exe* to begin. Linux users will download a *tar.gz* file. Processing can then be expanded in the terminal with `tar xvfz processing-xxxx.tgz` (replacing xxxx with the rest of the downloaded file's name, which is the version number). This will create a folder named something like *processing-1.0*. To start the program, change to that directory with `cd processing-xxxx` and run it with `./processing`. If you run into any problems, check the troubleshooting page at *http://wiki.processing.org/w/Troubleshooting* for help.

The Processing Interactive Development Environment (IDE) is very easy to use, and has similar controls to the Arduino IDE. In fact, the Arduino development environment was based on Processing's and uses many of the same concepts, with one very important difference. Arduino is fundamentally a C/C++ environment, while Processing's underlying language is Java. So while the code syntax you use will look similar at first, be aware that there are distinct differences in the commands and structure. One good way to get to know Processing is by learning about it in the online Getting Started guide (*http://processing.org/learning/gettingstarted*). The language itself is fully documented at *http://processing.org/reference/* (see Figure 5-9). There's also a comprehensive list of books at *http://processing.org/learning/books/*, including *Getting Started with Processing* (*http://oreilly.com/catalog/0636920000570/*) by Casey Reas and Ben Fry (O'Reilly) and *Learning Processing* by Daniel Shiffman (Kaufmann).

The Processing window has the same basic structure as Arduino's (see Figure 5-10). Programs are referred to as *sketches*. Buttons at the top allow the user to Run these sketches and Stop them; create New sketches; Open existing ones; Save them; and Export sketches as web applets. The center area is where code is edited, and the bottom of the window has a small gray space for messages, and for console output from the sketch. Sketches display visual output in a separate window.

Figure 5-8. The Download page on the Processing.org website

Figure 5-9. Processing's syntax and commands are fully documented on the website

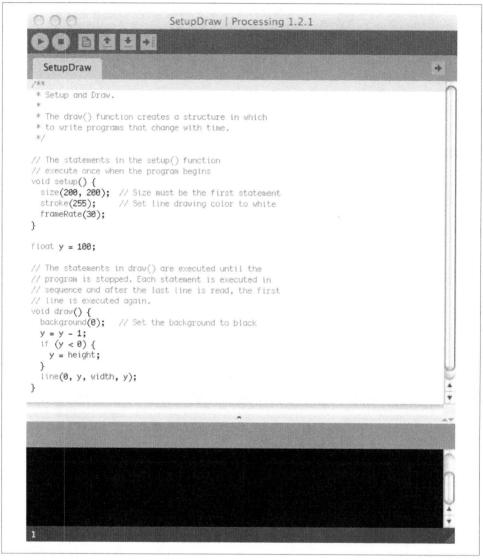

Figure 5-10. The Processing IDE with control buttons at the top, text area in the middle, and space for messages and console output at the bottom

Program the Base Station

The simple sensor network base station uses the following Processing program. Download the ZIP file of all the libraries and resources from this book's website. Inside the Processing sketch folder for the Simple Sensor Network program are two subdirectories called *code* and *data* (see Figure 5-11). The *code* folder contains the *log4j.jar* and *xbee-api-0.5.5.jar* library files. These contain all the code for communication with the

XBee in API mode. The *data* folder holds the *log4j.properties* file, required by *log4j.jar*. It also has a font file for a sans serif 10-point font used for screen display.

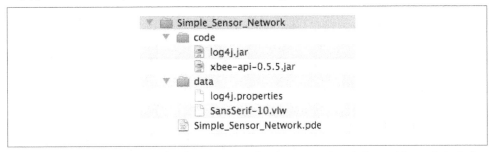

Figure 5-11. Directory structure for the Processing sketch program Simple Sensor Network, including all required libraries, config files, a font file, and the Processing ".pde" sketch itself

 You *must* replace the COM port listed in this code with your actual COM port. Look for it in the code around line 20. Port names are listed in the console in Processing, as your program starts up.

Once you have loaded the files and directories onto your computer and opened the *Simple_Sensor_Network.pde* file in Processing, press the Run button (labeled with a triangle) to launch the display code. It will open in a new window and show a thermometer for each sensor node detected, as shown in Figure 5-12.

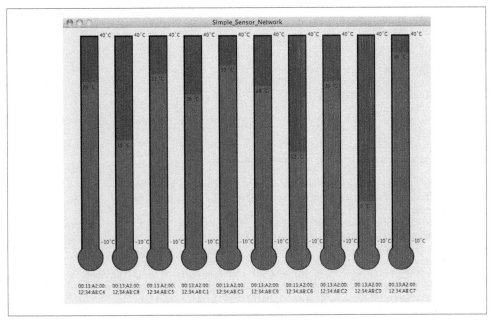

Figure 5-12. Simple Sensor Network temperature display screen in Processing

Simple Sensor Network display code in Processing

Here's the source code for the Processing sketch. The comment shown in bold about the serial port highlights an essential change. Other commented instructions are only important if you didn't download the source from the website listed in the Preface:

```
/*
 * Draws a set of thermometers for incoming XBee Sensor data
 * by Rob Faludi http://faludi.com
 */

// used for communication via xbee api
import processing.serial.*;

// xbee api libraries available at http://code.google.com/p/xbee-api/
// Download the zip file, extract it, and copy the xbee-api jar file
// and the log4j.jar file (located in the lib folder) inside a "code"
// folder under this Processing sketch's folder (save this sketch, then
// click the Sketch menu and choose Show Sketch Folder).
import com.rapplogic.xbee.api.ApiId;
import com.rapplogic.xbee.api.PacketListener;
import com.rapplogic.xbee.api.XBee;
import com.rapplogic.xbee.api.XBeeResponse;
import com.rapplogic.xbee.api.zigbee.ZNetRxIoSampleResponse;

String version = "1.01";

// *** REPLACE WITH THE SERIAL PORT (COM PORT) FOR YOUR LOCAL XBEE ***
String mySerialPort = "/dev/tty.usbserial-A1000iMG";

// create and initialize a new xbee object
XBee xbee = new XBee();


// make an array list of thermometer objects for display
ArrayList thermometers = new ArrayList();
// create a font for display
PFont font;

void setup() {
  size(800, 600); // screen size
  smooth(); // anti-aliasing for graphic display

  // You'll need to generate a font before you can run this sketch.
  // Click the Tools menu and choose Create Font. Click Sans Serif,
  // choose a size of 10, and click OK.
  font = loadFont("SansSerif-10.vlw");
  textFont(font); // use the font for text

  // The log4j.properties file is required by the xbee api library, and
  // needs to be in your data folder. You can find this file in the xbee
  // api library you downloaded earlier
  PropertyConfigurator.configure(dataPath("")+"log4j.properties");
  // Print a list in case the selected one doesn't work out
  println("Available serial ports:");
```

```
    println(Serial.list());
    try {
      // opens your serial port defined above, at 9600 baud
      xbee.open(mySerialPort, 9600);
    }
    catch (XBeeException e) {
      println("** Error opening XBee port: " + e + " **");
      println("Is your XBee plugged in to your computer?");
      println("Did you set your COM port in the code near line 20?");
    }
  }


  // draw loop executes continuously
  void draw() {
    background(224); // draw a light gray background
    SensorData data = new SensorData(); // create a data object
    data = getData(); // put data into the data object
    //data = getSimulatedData(); // uncomment this to use random data for testing

    // check that actual data came in:
    if (data.value >=0 && data.address != null) {

      // check to see if a thermometer object already exists for this sensor
      int i;
      boolean foundIt = false;
      for (i=0; i <thermometers.size(); i++) {
        if ( ((Thermometer) thermometers.get(i)).address.equals(data.address) ) {
          foundIt = true;
          break;
        }
      }

      // process the data value into a Celsius temperature reading for
      // LM335 with a 1/3 voltage divider
      //    (value as a ratio of 1023 times max ADC voltage times
      //     3 (voltage divider value) divided by 10mV per degree
      //    minus zero Celsius in Kelvin)
      float temperatureCelsius = (data.value/1023.0*1.2*3.0*100)-273.15;
      println(" temp: " + round(temperatureCelsius) + "°C");

      // update the thermometer if it exists, otherwise create a new one
      if (foundIt) {
        ((Thermometer) thermometers.get(i)).temp = temperatureCelsius;
      }
      else if (thermometers.size() < 10) {
        thermometers.add(new Thermometer(data.address,35,450,
                          (thermometers.size()) * 75 + 40, 20));
        ((Thermometer) thermometers.get(i)).temp = temperatureCelsius;
      }

      // draw the thermometers on the screen
      for (int j =0; j<thermometers.size(); j++) {
        ((Thermometer) thermometers.get(j)).render();
      }
```

```
    }
  } // end of draw loop


// defines the data object
class SensorData {
  int value;
  String address;
}


// defines the thermometer objects
class Thermometer {
  int sizeX, sizeY, posX, posY;
  int maxTemp = 40; // max of scale in degrees Celsius
  int minTemp = -10; // min of scale in degrees Celsius
  float temp; // stores the temperature locally
  String address; // stores the address locally

  Thermometer(String _address, int _sizeX, int _sizeY,
              int _posX, int _posY) { // initialize thermometer object
    address = _address;
    sizeX = _sizeX;
    sizeY = _sizeY;
    posX = _posX;
    posY = _posY;
  }

  void render() { // draw thermometer on screen
    noStroke(); // remove shape edges
    ellipseMode(CENTER); // center bulb
    float bulbSize = sizeX + (sizeX * 0.5); // determine bulb size
    int stemSize = 30; // stem augments fixed red bulb
                       // to help separate it from moving mercury
    // limit display to range
    float displayTemp = round( temp);
    if (temp > maxTemp) {
      displayTemp = maxTemp + 1;

    }
    if ((int)temp < minTemp) {
      displayTemp = minTemp;
    }
    // size for variable red area:
    float mercury = ( 1 - ( (displayTemp-minTemp) / (maxTemp-minTemp) ));
    // draw edges of objects in black
    fill(0);
    rect(posX-3,posY-3,sizeX+5,sizeY+5);
    ellipse(posX+sizeX/2,posY+sizeY+stemSize, bulbSize+4,bulbSize+4);
    rect(posX-3, posY+sizeY, sizeX+5,stemSize+5);
    // draw gray mercury background
    fill(64);
    rect(posX,posY,sizeX,sizeY);
    // draw red areas
    fill(255,16,16);
```

```
    // draw mercury area:
    rect(posX,posY+(sizeY * mercury),
        sizeX, sizeY-(sizeY * mercury)));

    // draw stem area:
    rect(posX, posY+sizeY, sizeX,stemSize);

    // draw red bulb:
    ellipse(posX+sizeX/2,posY+sizeY + stemSize, bulbSize,bulbSize);

    // show text
    textAlign(LEFT);
    fill(0);
    textSize(10);

    // show sensor address:
    text(address, posX-10, posY + sizeY + bulbSize + stemSize + 4, 65, 40);

    // show maximum temperature:
    text(maxTemp + "°C", posX+sizeX + 5, posY);

    // show minimum temperature:
    text(minTemp + "°C", posX+sizeX + 5, posY + sizeY);

    // show temperature:
    text(round(temp) + " °C", posX+2 ,posY+(sizeY * mercury+ 14));
  }
}

// used only if getSimulatedData is uncommented in draw loop
//
SensorData getSimulatedData() {
  SensorData data = new SensorData();
  int value = int(random(750,890));
  String address = "00:13:A2:00:12:34:AB:C" + str( round(random(0,9)) );
  data.value = value;
  data.address = address;
  delay(200);
  return data;
}

// queries the XBee for incoming I/O data frames
// and parses them into a data object
SensorData getData() {

  SensorData data = new SensorData();
  int value = -1;     // returns an impossible value if there's an error
  String address = ""; // returns a null value if there's an error

  try {
    // we wait here until a packet is received.
    XBeeResponse response = xbee.getResponse();
    // uncomment next line for additional debugging information
    //println("Received response " + response.toString());
```

```
      // check that this frame is a valid I/O sample, then parse it as such
      if (response.getApiId() == ApiId.ZNET_IO_SAMPLE_RESPONSE
          && !response.isError()) {
        ZNetRxIoSampleResponse ioSample =
          (ZNetRxIoSampleResponse)(XBeeResponse) response;

        // get the sender's 64-bit address
        int[] addressArray = ioSample.getRemoteAddress64().getAddress();
        // parse the address int array into a formatted string
        String[] hexAddress = new String[addressArray.length];
        for (int i=0; i<addressArray.length;i++) {
          // format each address byte with leading zeros:
          hexAddress[i] = String.format("%02x", addressArray[i]);
        }

        // join the array together with colons for readability:
        String senderAddress = join(hexAddress, ":");
        print("Sender address: " + senderAddress);
        data.address = senderAddress;
        // get the value of the first input pin
        value = ioSample.getAnalog0();
        print(" analog value: " + value );
        data.value = value;
      }
      else if (!response.isError()) {
        println("Got error in data frame");
      }
      else {
        println("Got non-i/o data frame");
      }
    }
    catch (XBeeException e) {
      println("Error receiving response: " + e);
    }
    return data; // sends the data back to the calling function
  }
```

Troubleshooting

If things don't work at first, here are some steps to try:

1. Check all your electrical connections to make sure there are no loose wires and that all the components are connected properly.

2. Check the coordinator configuration in X-CTU again, including that the correct modem type (XB24-ZB) and function set (ZigBee Coordinator API) have been selected. Make sure that ATAP has been set to 2 for this project! Also check that the PAN ID is configured as you expect.

3. Check the router configuration in X-CTU to confirm that the correct modem type (XB24-ZB) and function set (ZigBee Router AT) have been selected. Also check that the PAN ID, destination high, and destination low are configured as you expect, and that ATJV, ATD0, and ATIR have been configured as described above.

4. An LED placed from the ASSOC pin of each sensor XBee (physical pin 15) to ground should show a flashing light.

5. If your serial adapter has an RSSI light, it should illuminate when the radio is receiving information. If messages stop coming in, this light will time out and go dark after 10 seconds.

6. Use a multimeter to see if the voltage at the D0 pin of each sensor XBee (physical pin 20) varies with changes in the temperature. It should be somewhere in the range between 0 and 1.2 volts and change as you warm or cool the LM335.

7. If your temperature readings are somewhat off, you can calibrate the sensor by connecting a potentiometer across the LM335 with the output connected to the adjustment pin as in Figure 5-13. Also, check to see if it is next to another component on your circuit board that's generating a bit of heat, like the voltage regulator.

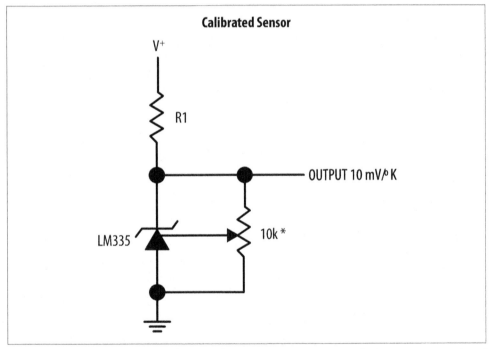

Figure 5-13. LM335 calibration schematic showing a potentiometer attached to power, ground, and the adjustment pin

8. We are not always able to see our own mistakes. Have a friend check everything for you. Sometimes only a second pair of eyes will catch the one or more issues that are standing in the way of success.

9. When all else fails, try taking a break and coming back to the project after a good night's rest.

Yay!

You have learned a lot and built your very first complete wireless sensor network. Congratulations! While you won't always need to worry about the API mode, having some understanding of it will help you both in your work with the XBees and in general as new wireless communications protocols are developed and dispensed to the networking community. In the next chapter we will dial back the intensity a little to show you some useful power management tricks, and build a second network that allows you to control household appliances. Before that, take a moment to revel in securing your official Sensor Networking Merit Badge. You've earned it!

CHAPTER 6

Sleeping, Then Changing the World

Whew! At this point, the book has covered a lot of ground and you have come a long way in a very short time. Now we'll ease the pace just a bit and address some nuances of ZigBee mesh networking, including sleep mode, end devices, and power management. Then it's time to change things in the world with direct actuation. This chapter features a powerful control project you could use to automate your home or to play Pong using the window lights in a skyscraper.

Sleep Mode

Going wireless often means cutting the electric cord. Projects that are mobile or remotely located frequently use batteries or another constrained power source that demands economizing on energy. XBee radios, like many other communications and microcontroller devices, can put themselves into a temporary sleep state where nearly no current is consumed. The trade-off is that during this state no activities can take place. During sleep, the device is almost completely turned off and is incapable of receiving or sending messages until it wakes back up. ZigBee mesh networking is specifically designed to smoothly handle communications on a network where at any moment many radios might be in this type of low-power state. In fact, by getting very stingy and using the right kind of power cell, we can envision ZigBee networks where batteries last many years at a time, while the networks still perform sufficient sensing and dependable actuation.

End Devices

We introduced *end devices* in Chapter 2. You may recall that end devices are essentially stripped-down versions of router radios. They can join networks and participate in communications, but because they power down intermittently, they explicitly do not act as messengers between any other devices. End devices always require a router or the coordinator to be their parent device. The end devices' parent helps them join the network to begin with and then stores messages for them when they are asleep. ZigBee networks may have any number of end devices. It's perfectly legitimate to create a network with a single coordinator acting as the sole parent for many end devices.

Storing and forwarding

When an end device is asleep, it is pretty much dead to the world. Any attempt to contact it by radio will fail because during hibernation the end device turns off its transmitter and receiver to conserve energy. That's why it needs a parent. One of the parent's jobs is to act as its mailbox, storing messages while the end device is asleep and forwarding them when the end device wakes back up. The portion of the protocol enabling network communications with sleeping end devices is automatically managed right inside the mesh network radios. ZigBee networks get this feature without any additional components or code required to manage the process.

Constraints

Alas, the power-saving lunch does not come entirely for free. The longer a node sleeps, the longer any message to it must be stored. XBee routers that act as the parent to end devices generally store only a single message, so communications must be carefully designed with those limitations in mind. Network chatter must be kept to a minimum to avoid overwhelming the parent device's storage resources and discarding important information. Also, radios configured as end devices are by definition incapable of acting as mesh routers themselves. After all, while the radio is asleep it is useless for retransmitting data and therefore presents a virtual cul de sac to the mesh network. On the upside, considerable power is conserved by forgoing these responsiveness and range extension features. You will want to carefully consider these trade-offs when designing your networks.

 An XBee router must reserve memory to store mailbox data for each of its end device children. Usually the number of child nodes per router is limited to about 8 or 10. If you like, you can check on how many remaining children a given router or coordinator will support with the ATNC command.

Advantages

Using sleep mode, an end device can stretch battery life from hours into days, weeks, and sometimes even years. For example, if full-time use of a (hypothetical) battery would drain it in four hours, putting the radio into a cycle where it slept for one second, then woke for one second before sleeping again could (roughly) double that battery's life to eight hours. Cyclically sleeping for 59 seconds and then waking for a single second might keep the same battery going for something on the order of 10 days. Taking this further, sleeping for the same 59.98 seconds and then only waking up for 20 milliseconds would extend the available power to over a year. Sleep mode can have tremendous benefits. Keep in mind, though, that real-world battery life predictions aren't quite so simple.

 Batteries are physical, chemically active devices. Every type of battery chemistry has its own complex characteristics that bedevil any attempt to make simplistic calculations on power use. Factors like temperature, charge state, instantaneous draw levels, aging timetables, self-draining curves, and materials purity all weigh into the complex nonlinear functions required to properly estimate battery life. Nevertheless, sometimes simplistic estimations are all that's needed to select an appropriate sleep state that works within the context of your project. Build an ample safety factor into your back-of-the-napkin battery arithmetic and you'll find you can get a lot of reasonable estimations accomplished without getting mired in the awesome complexities of chemical engineering.

Configuring Sleep

There are six AT commands associated with sleeping (see Table 6-1). These commands work together to configure the specific behaviors that are most appropriate for your particular project. In many cases, you'll only really need to set three of them: Sleep Mode, Sleep Period, and Time Before Sleep. The other three commands are for more unusual configurations so they are less commonly needed.

Sleep Mode

The XBee ZigBee End Device radios have four basic sleep behaviors. These are set with the ATSM command:

ATSM 0: *Disables sleep mode*
    The radio will always be awake and using power, but because it is running the end device firmware it will not route for other radios and it still requires a parent device.

ATSM 1: *Pin wake mode*
    In this mode the XBee module will sleep when its sleep control pin—physical pin 9—is asserted or pulled high by connecting it to 3.3 volts (Figure 6-1). The radio will finish transmitting or receiving before it goes to sleep. When the sleep control pin is brought low by connecting it to ground, the radio will wake from sleep and be able to transmit and receive again. When asleep in this mode, the radio uses less than 10 microamps, a minuscule amount of power. Waking it up takes about 13 milliseconds. You use pin wake mode when there's another device—most commonly some kind of microcontroller—available to assert and de-assert the signal going to pin 9.

ATSM 2 *and* 3
    These modes are currently undefined.

On/
Sleep
(output)

Sleep
control
(input)

Figure 6-1. XBee sleep control pin 9 accepts external input for putting the module to sleep. The On/
Sleep pin 13 goes low when the radio sleeps and is brought high while the XBee is awake.

ATSM 4: *Cyclic sleep mode*

When an XBee radio is running independently and there is no other device to switch it in and out of sleep mode, it still has the internal capacity to sleep and wake on a fixed schedule. This is the most common way of conserving power when a radio is being used as a simple sensor node. You set how often the device sleeps and for how long using ATSP and ATST as described below. Upon wake-up, the end device will poll its parent to see if there are any incoming messages waiting for it. When asleep in cyclic mode, the radio uses less than 50 microamps, a very modest amount of power. Waking it up takes only 2 milliseconds.

ATSM 5: *Cyclic sleep with pin wake*

This is basically the same as regular cyclic sleep mode but with the option of also waking the module using physical pin 9.

Sleep Period

The length of time an XBee radio remains asleep is set with the ATSP command. The sleep period setting gets multiplied by 10, with the result being the number of milliseconds of sleep time. SP will accept any setting in the hexadecimal range for 0x20 to 0xAF0. When multiplied by 10, this means that the basic sleep period can be as short as 320 milliseconds and as long as 28 seconds. The sleep period can effectively be extended by setting the number of consecutive sleeps with ATSN and ATSO (as described below).

The coordinator and router radios in an interactive network should have their SP register set to the same or greater value as the end devices. On parent nodes, SP defines how long to store messages in the forwarding buffer before discarding them. This way parents won't throw away messages before the child end-nodes have a chance to wake up and retrieve them.

Time Before Sleep

With the ATST command, you can set the minimum timeout for the radio to remain awake before returning into cyclic sleep. The XBee module will never go into its low power mode while a message is being transmitted or received, so think of this timeout as a period of silence that is required before the radio can fall asleep again. Time before sleep can be set to as little as 0x1 for one millisecond of timeout, or as high as 0xFFFE for about 65 seconds of delay before returning to low-power cyclic sleep.

Advanced commands

Several more sleep commands go beyond the basics. These are often used when an XBee radio is working with an external device, and controlling it with the XBee's internal timers. Imagine a weather station that might need to warm up for a short time before it is ready to supply a data sample. Alternately, imagine an application that uses the radio to remotely wake and activate a traffic information sign. If there's no incoming information, it's not necessary to wake the sign and consume valuable battery power for no reason. The following commands are intended to help in these special situations, and can also be employed in cases where a particularly lengthy sleep time is desired.

The On/Sleep pin, physical pin 13 on the XBee, goes high when the module is awake and low when it is sleeping. Attaching an LED to this pin gives a visual indication of the radio's current sleep state. This pin can also be used to control an external device—for example, so that it is powered up only when the radio is awake. (See Figure 6-1.)

ATSN: *Number of Sleep Periods*
This command specifies how many sleep periods to *skip* asserting the On/Sleep pin if no data is received on wake-up. Setting it to the default of 0x01 causes the On/Sleep pin to be asserted on every wake-up. Setting it higher—for example, to 0x09—would allow eight wake-ups with no incoming data to pass before asserting the On/Sleep pin. The maximum value for this register is 0xFFFF or 65,535 wake-up checks before the On/Sleep pin is forcibly asserted. Remember that if incoming data is received during *any* wake-up, the On/Sleep pin will be brought high, no matter how ATSN is set. A secondary use of this register is as a multiplier for ATSP, when you need the radio to sleep for very long times. See ATSO below.

ATWH: *Wake Host*

In some applications, a sensor or device might need time to be turned on, boot up, and stabilize before the XBee either sends received messages out of its local serial port or samples its local pins for I/O data. In these cases it's helpful to have a specific delay after the On/Sleep pin is asserted to turn on the attached device. The wake host delay can be as little as the default of 0x0 for no delay, or as long as 0xFFFF for about 65 seconds of waiting time before communication or I/O sampling.

ATSO: *Sleep Options*

There are currently three states for this register. 0x0 is the default setting of no options enabled. 0x02 tells the radio to always wake up for the entire ST time, even when no data is waiting to be sent or received. This is useful only in specialized external device control situations. The 0x04 option setting forces the radio to sleep continuously for the entire period specified by SN * SP, to a maximum total time of 1,834,980,000 milliseconds or just over three weeks between wake-ups. This last option has the potential to allow the right battery to last for many years!

> The complete formula for calculating sleep time when you are using the advanced options is SP * 10 * SN, when SO is set to 0x04.

Table 6-1. Summary of AT commands for sleeping the XBee radios

AT command	Name and description	Node type	Parameter range	Default
SM	**Sleep Mode.** Sets the sleep mode on the RF module.	E	0 - Sleep disabled 1 - Pin sleep enabled 4 - Cyclic sleep enabled 5 - Cyclic sleep, pin wake	0
SN	**Number of Sleep Periods.** Sets the number of sleep periods to not assert the On/Sleep pin on wake-up if no RF data is waiting for the end device. This command allows a host application to sleep for an extended time if no RF data is present.	CRE	1–0xFFFF	1
SP	**Sleep Period.** Determines how long the end device will sleep at a time, up to 28 seconds. (The sleep time can selectively be extended past 28 seconds using the SN command.) On the parent, this value determines how long the parent will buffer messages for the sleeping end device. It should be set at least equal to the longest SP time of any child end device.	CRE	0x20–0xAF0 (×10 ms) (Quarter-second resolution)	0x20
ST	**Time Before Sleep.** Sets the time-before-sleep timer on an end device. The timer is reset each time serial or RF data is	E	1–0xFFFE (×1 ms)	0x1388 (5 seconds)

AT command	Name and description	Node type	Parameter range	Default
	received. Once the timer expires, an end device may enter low-power operation. Applicable for cyclic sleep end devices only.			
SO	**Sleep Options**. Configures options for sleep. Unused option bits should be set to 0. Sleep options include: 0X02 – Always wake for ST time 0x04 – Sleep entire SN * SP time Sleep options should not be used for most applications.	E	0–0xFF	0
WH	**Wake Host**. Sets or reads the wake host timer value. If the wake host timer is set to a nonzero value, this timer specifies a time (in millisecond units) that the device should allow after waking from sleep before sending data out the UART or transmitting an I/O sample. If serial characters are received, the WH timer is stopped immediately.	E	0–0xFFFF (× 1 ms)	

Easy Sleeping

There are a lot of sleep options but the good news is that you only have to set them once for your application and, in most cases, you only need one or two settings. For instance, if we just want the XBee module to wake up briefly every five seconds, selecting ATSM 4, and ATSP 1F4 turns on cyclic sleep mode and sets the period of time the radio is asleep to 5000 milliseconds. Remember that all the commands use hexadecimals and that the SP register is always multiplied by 10. So the hexadecimal 0x1F4 translates to 500 in decimal, and when multiplied by 10 results in 5,000 milliseconds, or 5 seconds.

 Waking up from sleep *always* triggers an I/O sample, as long as ATIR is set to a nonzero number and at least one pin is configured as a digital or analog input. Samples will continue at the IR rate until the ST timer has expired, and then the radio will sleep again.

Simple Sensor with Sleep Project

The simple sensor network from the previous chapter is a prime candidate for some power-saving assistance from sleep mode. You'll use the same base station configuration, but you can either add new end nodes to the network or replace your existing hardwired nodes with battery-powered ones. The following instructions are for a single sleeping node. Create as many of these as you like; just remember the base station still needs to be a coordinator radio, so its configuration should remain the same.

The parts listed below supply enough battery voltage to continue using the voltage regulator circuits you already built. Alternatively, you could remove the voltage regulation and use two AA batteries to power the end nodes.

Parts

- 9-volt batteries (RS 23-866)
- 9-volt to barrel jack adapters (SFE PRT-09518)

...or

- AA batteries (RS 23-942)
- 4xAA to barrel jack connectors (SFE PRT-09835)

...and

- XBee radios (Series 2/ZB firmware) configured as a ZigBee End Device AT mode (Digi: XB24-Z7WIT-004, DK 602-1098-ND)

Prepare Your End Device Radios

Follow the instructions under "Reading Current Firmware and Configuration" on page 35 in Chapter 2 to configure each of your sensor node radios as a ZigBee End Device AT.

 Your *end device* radios will use the *AT* firmware so you can easily configure them using a serial terminal. Be sure you select the AT version for your end devices!

Each end device radio can be labeled with an "E."

Configure Your End Device XBees

We'll continue to use CoolTerm and an XBee Explorer USB adapter to set up the radios. For each of your sleeping end device sensor node radios:

1. Select an end device XBee you've labeled with an "E" and place it into the XBee Explorer.
2. Plug the XBee Explorer into your computer.
3. Run the CoolTerm program and press the Options button to configure it.
4. Select the appropriate serial port and check the Local Echo box so you can see your commands as you type them.
5. Click on the Connect button to connect to the serial port.

6. Type **+++** to go into command mode. You should receive an OK reply from the radio.

7. Select the *same* PAN ID that you entered for your original simple sensor network.

8. Type **ATID** followed by the PAN ID you selected and press Enter on the keyboard. You should receive OK again as a reply.

9. Every ZigBee coordinator always has 0 as its 16-bit network address. Type **ATDH 0** and press Enter on the keyboard. You should receive an OK response.

10. Enter **ATDL** followed by the *low* part of your radio's *destination* address (0, the address of the coordinator). Type **ATDL 0** and press Enter. You should receive an OK response.

11. Enter **ATD02** to put pin 0 in analog mode.

12. Enter **ATIR3E8** to set the sample rate to 1,000 milliseconds.

13. Enter **ATSM4** to put the radio into cyclic sleep mode.

14. Enter **ATSP64** to sleep the radio for one second (100 * 10 ms).

15. Enter **ATST14** to time out and sleep 20 milliseconds after each sample transmission.

16. Save your new settings as the radio's default by typing **ATWR** and pressing Enter.

 Once you put your radio into sleep mode it may appear to become unresponsive. That's because it is sleeping! Don't panic if you can't wake it up. See the sidebar "Wake-Up Issues and Reset Strategies" on page 170 for more information on waking a sleeping radio.

Add sensor nodes...

1. If you've decided to add *new* temperature sensor nodes, make them exactly like those in the previous chapter. The radios you just configured can be plugged right into these new sensor boards.

2. Power your new sensor boards with a battery pack to make them totally untethered.

...or replace sensor nodes

1. If you don't want to add any new sensor nodes, you can also replace the radios in your existing sensor boards with the ones you just programmed for sleep mode.

2. Power your newly sleeping sensor boards with a battery pack for full mobility.

3. Run the Simple Sensor Network program in Processing to test your new and/or replaced sensor nodes.

Figure 6-2 illustrates the simple sensor network with end devices.

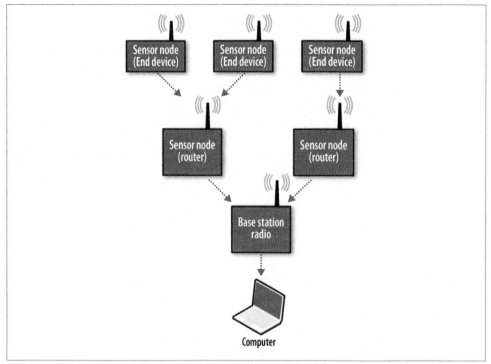

Figure 6-2. Simple sensor network with sleeping end devices

 In the Simple Sensor with Sleep project, it is *not* necessary to add the SP setting to the parent routers or coordinator. This is because they won't be storing any outgoing messages so the buffer timeout won't ever be used. However, it doesn't hurt and it is a good habit to have. Set ATSP to 1F4 on every device in the network, but only if you want to.

Wake-Up Issues and Reset Strategies

Once you set ATSM (sleep mode) to something other than zero, there will be times when the radio refuses to respond at all. This is because it is sleeping! Don't worry; you can still wake the radio up to talk to it. Here are some strategies to rouse your snoozing radio for reconfiguration:

- If the radio is in sleep mode 1 (pin sleep) or sleep mode 5 (cyclic sleep with pin wake-up), bringing physical pin 9 low by connecting it to ground will wake the radio.

- For radios in sleep mode 4, there are a few options. Assuming the radio is using AT command firmware, you could keep issuing the **+++** sequence with a one-second or longer pause in between each attempt. Sooner or later the radio will respond with an OK and you can take it out of sleep mode (ATSM0) or issue other commands. If you've set the sleep period (ATSP) to be very long or the time before sleep (ATST) to be very short, it may take quite some time to get lucky enough to issue the command just as the radio wakes up. Be patient and don't forget to pause at least one second between **+++** attempts.

 If you are using a full-featured serial adapter, you may be able to see the CTS indicator in your terminal program light up when the radio is awake. Both CoolTerm and X-CTU have this indicator. Time your **+++** attempts to happen when CTS is active for fastest results.

- X-CTU is your wake-up buddy. If the radio is sleeping when you attempt to configure it with X-CTU, X-CTU will show a dialog box (refer back to Figure 2-11) that suggests resetting the radio. If your adapter board doesn't have a reset button, carefully lifting the radio out of its serial adapter sockets (yes, while the socket is still plugged into USB) and reseating it will effectively reset it and should wake it up.

- If all else fails, you can always force-download new firmware to the radio. This type of drastic step is usually not necessary. On the PC Settings tab, set Flow Control to Hardware. On the Modem Configuration tab, check the Always Update Firmware checkbox. In the Modem pop-up list, select XB24-ZB along with the function set you desire, such as ZigBee End Device AT. Press the Show Defaults button and then the Write button. You may be asked to reset the radio, in which case carefully unseat it and reseat it in its serial adapter sockets. It may take several tries to get the firmware to reload, but patience tends to pay off here.

- When all else fails, don't forget that Digi has excellent tech support staff that can probably help you resuscitate your radio (*http://digi.com/support*).

Direct Actuation

Creating sensor networks is a lot of fun and tremendously useful. There are plenty of reasons to collect data from multiple nodes and bring it to a central location. There are equally great reasons to take commands from a central location and create real events in multiple physical locations remotely. The XBee radio is capable of receiving commands that set its digital I/O output pins to trigger real-world events without the use of any external microcontroller. By itself, the XBee can power an LED, sound a small buzzer, or even operate a tiny motor. Many more devices can be operated directly from the XBee with the use of a relay. Relays are really just electrically operated switches. They allow low currents to turn on and off devices that require much higher currents to operate. Relay-type devices for our purposes include transistors that can transform low current outputs into medium ones, as well as the larger mechanical and electrical relays that typically use medium currents to switch high-power loads, such as large DC motors or even wall-plug A/C appliances. The example at the end of this chapter will have you turning on and off home electronics quite reliably with just an XBee and a set of relays doing all the actuation work.

Naturally, not every application that wants to have real-world effects can forgo an external microcontroller. Many devices will benefit from additional local decision-making. For example, a lamp that can be turned on remotely may behave more intelligently if it first checks a photocell to see whether the room is already flooded with daylight. Many applications also require more digital outputs than the 10 that XBee radios provide. A simple 12-segment bar graph is going to need a microcontroller to provide the required outputs to drive its display. Don't be afraid to use an external microcontroller if it makes your project easier to prototype, or of course if it is required to enable basic functionality.

Direct Actuation Example

Remote control impresses people quite a lot, even though it isn't terribly difficult. This project will allow you to control lamps and other small home appliances wirelessly from your computer. It can serve as the basis for any number of interesting control systems, from basic automated home lighting to that complex interactive robotic opera that's been kicking around in your head the past few years. Or you could create Pong on the side of a building (*http://blinkenlights.net/blinkenlights*). A wireless switch is a wonderful thing.

This project has two remote nodes. It can be completed with just a single remote node if you are on a tight budget.

Parts

- Two solderless breadboards (AF 64, DK 438-1045-ND, SFE PRT-09567)
- Hookup wire or jumper wire kit (AF 153, DK 923351-ND, SFE PRT-00124)
- Two 9-volt or 5-volt power supplies (9-volt batteries also work well for short-term use) (AF 63 or 80 with 9 V battery, RS 273-355, SFE TOL-08269 or TOL-00298)
- Two 3.3 V voltage regulators (TO-220 package) (DK 497-1491-5-ND, SFE COM-00526)
- Two DC power jacks (2.1 mm ID, 5.5 mm OD) (DK CP-024A-ND, RS 274-1577, SFE PRT-00119)
- Assorted 5 mm LEDs (DK 160-1707-ND, RS 276-041, SFE COM-09590)
- Two 10K ohm resistors (DK P10KBACT-ND, SFE COM-08374)
- Two 10 µF electrolytic capacitors (DK P966-ND, RS 272-1025, SFE COM-00523)
- Two 1 µF electrolytic capacitors (DK P993-ND, RS 272-1434)

- Two 2N3904 transistors (DK 2N3904TFCT-ND, RS276-2016, SFE COM-00521)
- Two PowerSwitch Tails, 5-volt relay for A/C loads (*http://powerswitchtail.com* or SFE COM-09842)
- One XBee radio (Series 2/ZB firmware) configured as a ZigBee Coordinator API mode (Digi: XB24-Z7WIT-004, DK 602-1098-ND)
- Two XBee radios (Series 2/ZB firmware) configured as a ZigBee Router AT mode (Digi: XB24-Z7WIT-004, DK 602-1098-ND)
- Two XBee breakout boards with male headers and 2 mm female headers installed (AF 126 [add SFE PRT-00116], SFE BOB-08276, PRT-08272, and PRT-00116)
- XBee USB serial adapter (XBee Explorer, Digi Evaluation board, or similar) (AF 247, SFE WRL-08687)
- USB cable for XBee adapter (AF 260, SFE CAB-00598)
- Wire strippers (AF 147, DK PAL70057-ND, SFE TOL-08696)
- Small screwdriver (RS 64-069, SFE TOL-09146)
- A lamp or any other small A/C appliance that draws less than 10 resistive amps

Prepare Your Coordinator Radio

1. Follow the instructions under "Reading Current Firmware and Configuration" on page 35 in Chapter 2 to configure one of your radios as a ZigBee Coordinator API.

 Your *coordinator* radio *must* use the API firmware for this project to work because I/O data is only delivered in API mode. Be sure to select the API version for your coordinator!

2. Use X-CTU to configure the coordinator with a PAN ID (between 0x0 and 0xFFFFFFFFFFFFFFFF) that you've selected. Write down this PAN ID so you can program your router radios with the same one. Every radio in your network must use the same PAN ID so that they can communicate with each other:

Pan ID:

3. The software libraries that we are using in Processing *require* that the base station XBee be in API Mode 2 (API Operation with escaped characters). Use X-CTU to set ATAP to 2, and Write the configuration to your radio.

4. Label the coordinator radio with a "C" so that later you'll know which one it is.

The XBee Java API Library communicates using escaped character mode, as described in a note under "Libraries" on page 141. Be sure that you set the coordinator's API to mode 2; otherwise the project will not work!

Prepare Your Router Radios

1. Follow the instructions under "Reading Current Firmware and Configuration" on page 35 in Chapter 2 to configure each of your actuator node radios as a ZigBee Router AT.

2. Your *router* radios will use the *AT* firmware so you can easily configure them using a serial terminal. Be sure you select the AT version for your routers!

3. Each router radio can be labeled with an "R."

Prepare the Actuator Boards

It's not a bad idea to use ATRE to reset your router radios to factory defaults if you are reusing them after another project. This way the radio won't have any weird legacy configurations lurking in its registers.

Configure Your Router XBees

We'll use the CoolTerm terminal program and an XBee Explorer USB adapter again to set up your radios. For each of your actuator node radios:

1. Select a router XBee you've labeled with an "R" and place it into the XBee Explorer.

2. Plug the XBee Explorer into your computer.

3. Run the CoolTerm program and press the Options button to configure it.

4. Select the appropriate serial port, and check the Local Echo box so you can see your commands as you type them.

5. Click on the Connect button to connect to the serial port.

6. Type +++ to go into command mode. You should receive an OK reply from the radio.

7. Select the **same** PAN ID you entered for your first radio above.

8. Type **ATID** followed by the PAN ID you selected and press Enter on the keyboard. You should receive OK again as a reply.

9. Every ZigBee coordinator always has 0 as its 16-bit network address. Type **ATDH 0** and press Enter on the keyboard. You should receive an OK response.

10. Enter **ATDL** followed by the *low* part of your radio's *destination* address, in this case also a zero because that's the fixed address for the coordinator. Type **ATDL 0** and press Enter. You should receive an OK response.

11. Enter **ATJV1** to ensure that your router attempts to rejoin the coordinator on startup.

12. Enter **ATDO4** to set pin 0 as low digital output to begin with.

13. Save your new settings as the radio's default by typing **ATWR** and pressing Enter.

 It's always good to recheck your configurations after you enter them. For example, to recheck that you entered the destination address correctly, from command mode, type **ATDL** and press Enter to see the current setting.

Connect voltage regulator circuit and power jack to breadboard

1. Wire up a breadboard with a 3.3-volt voltage regulator (LD1117V33) as shown. The regulator has three legs—typically, ground, output, and input—when viewed from the front (where the writing is). Sometimes these legs are in a different order, so find and check the data sheet if you're not sure! Input is where a high voltage, for example 5 or 9 volts, is applied to the regulator. Output is where you will get the regulated 3.3 volts. Ground is the common ground for your entire circuit, including input, output, and all the other components. Bring ground out to both blue ground rails that run along the sides of your breadboard. Bring 3.3-volt output power to both of the red power rails (refer back to Figure 5-4).

2. Solder a red wire (about 10 cm) to the short center pin of your power jack, and solder a similar black wire to the longer outer pin (refer back to Figure 5-5). Don't allow the two connections to touch each other since that will create a short circuit when you power up!

3. Attach the red wire from the power jack, using the breadboard to connect it to the *input* pin of the voltage regulator. Attach the black ground wire to the ground pin of the voltage regulator in the same way.

4. Hook up the *output* pin of the voltage regulator to one of the power rails of the breadboard using a red wire. Hook up the ground pin to one of the ground rails on the breadboard.

5. Use the two capacitors to "decouple" the power supply in the following way: attach the short ground lead of the 10 µF capacitor (also marked with a stripe on the capacitor's ground side) to ground near the voltage regulator. Attach the other positive lead of the 10 µF capacitor to the voltage regulator's *input* pin. This will remove some lower-frequency noise coming from the wall power supply. Also attach the short ground lead of the 1µF capacitor to ground, and the other positive lead to the 3.3 V *output* pin. This will remove some higher-frequency noise coming out of the voltage regulator. Decoupling will prevent noisy power from reaching your radio and interfering with its signal.

6. Hook up power and ground across the breadboard so that the rails on both sides are live.

 It's a really good idea to check the voltage levels using a multimeter after you first wire up the breadboard for power. Make sure that your power rails have 3.3 volts on both sides where you expect it. You don't want to send 9 volts to your radio and cook it!

Router XBee connection to power

1. With a *router* XBee mounted on its breakout board, position the breakout board in the center of your other breadboard so that the two rows of male header pins are inserted on opposite sides of the center trough.

2. Use red hookup wire to connect pin 1 (VCC) of the XBee to 3.3-volt regulated power.

3. Use black hookup wire to connect pin 10 (GND) of the XBee to ground.

Transistor and relay output

This project uses the PowerSwitch Tail A/C relay (see Figure 6-3). This relay is usually activated by 5 V of direct current. The XBee can't provide enough voltage or amperage by itself to drive that relay, so we use an NPN transistor as an electronic switch to send 5 volts directly to the relay. Think of it as a switch that throws another switch. You can get the data sheet for the 2N3904 transistor at *http://www.fairchildsemi.com/ds/2N %2F2N3904.pdf*, and the schematic for the PowerSwitch Tail at *http://powerswitchtail .com/Documents/PST%20Instructions%20v1.03.pdf*.

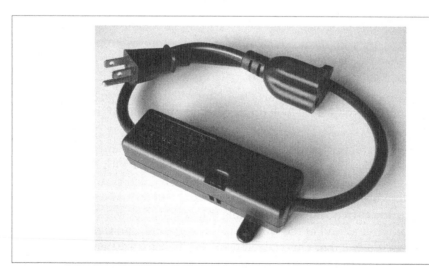

Figure 6-3. PowerSwitch Tail 5 V relay for A/C loads

1. The 2N3904 transistor has three leads. When the transistor's flat side is facing you, the leads from left to right are emitter, base, and collector. Insert the 2N3904 so that each lead is in its own row on the breadboard.

2. Use a black wire to connect the leftmost emitter lead to one of the ground rails.

3. Insert the 10K ohm resistor so that it is connected to XBee digital output 0 (physical pin 20) on one end and to the center base pin of the 2N3904 on the other. You can use jumper wires to make the connection if that's more convenient in your breadboard layout.

4. Connect a black wire from the rightmost collector pin of the 2N3904 to the negative screw terminal on the PowerSwitch Tail. It should be marked with a - (minus sign). Use a small screwdriver to tighten the screw terminal so that the black wire is securely attached.

5. Connect a red wire from the input voltage coming from your DC power jack (the same pin that feeds the input of your voltage regulator with 5 to 9 volts) to the positive screw terminal on the PowerSwitch Tail. That screw terminal should be marked with a +. Use a small screwdriver to tighten the screw so that the red wire is securely attached.

Second actuator board

Create the second actuator board in the same way as the first. You can make as many actuator systems as you like. The software will work with as few as one or as many as five without any adjustment to the software. Figure 6-4 shows the breadboard layout for the simple actuator node, and Figure 6-5 shows the schematic.

PowerSwitch Tail A/C relay

Plug one end of the PowerSwitch Tail into an A/C wall outlet. The other end can have a lamp or any similar appliance that draws less than 10 amps at 120 V plugged into it. See the data sheet for additional information (*http://powerswitchtail.com/Documents/PST%20Instructions%20v1.03.pdf*). You'll be able to turn this device on and off wirelessly, right from your computer using the on-screen switches displayed by the Processing program.

Prepare the Base Station

Your base station radio is simply an XBee serial adapter connected to your computer.

Connect to your computer

1. Select the coordinator XBee you've labeled with a "C" and place it into the XBee Explorer.

2. Plug the XBee Explorer into your computer.

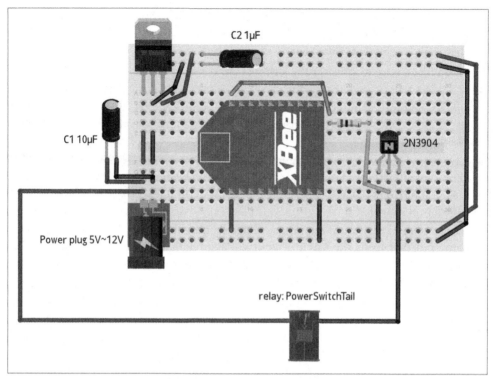

Figure 6-4. Simple actuator node breadboard layout

Program the actuator network base station

The simple actuator network base station uses the Processing program that follows. You can download a ZIP file of all the libraries and resources from this book's website. You can also find the XBee API library at *http://code.google.com/p/xbee-api/* and Processing at *http://processing.org/download* for Linux, Macintosh, or Windows.

Inside the Processing sketch folder for the Simple Actuator Network program are two subdirectories called *code* and *data* (see Figure 6-6). The *code* folder contains the *log4j.jar* and *xbee-api-0.5.5.jar* library files. These contain all the code for communication with the XBee in API mode. The *data* folder holds the *log4j.properties* file, required by *log4j.jar*. It also has a font file for a sans serif 10-point font used for screen display and two *.jpg* images for the on and off switch positions.

If you download the file from this book's website, simply unzip it and launch the *Simple_Actuator_Network.pde* file using Processing.

 Be sure to replace the COM port listed in this code with your actual COM port. Port names are listed in the console in Processing, as your program starts up.

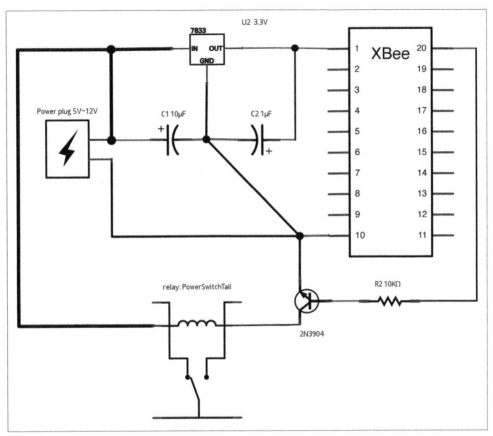

Figure 6-5. Simple actuator node schematic

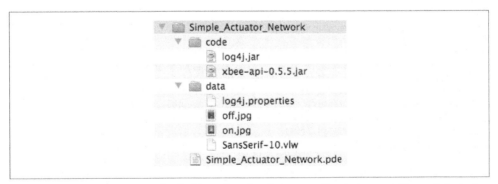

Figure 6-6. Directory structure for the Processing sketch program Simple Actuator Network, including all required libraries, config files, a font file, two image files, and the Processing ".pde" sketch itself

Once you have loaded the files and directories onto your computer and opened the *Simple_Actuator_Network.pde* file in Processing, press the Run button (labeled with a triangle) to launch the display code. It will open in a new window and show a switch for each actuator node detected, as shown in Figure 6-7.

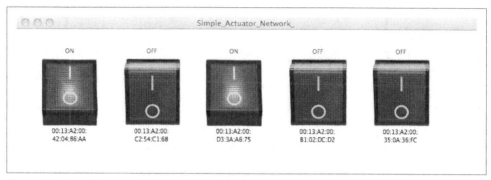

Figure 6-7. Simple Actuator Network switch display screen in Processing

Simple Actuator Node Code in Processing

Here's the source code for the Processing sketch. The comment shown in bold about the serial port highlights an essential change. Other commented instructions are only important if you didn't download the source from the website listed in the Preface (however, you'll still need to download this source code to obtain the *on.jpg* and *off.jpg* images used in this example):

```
/*
 * Draws a set of switches for managing XBee Actuators
 * by Rob Faludi http://faludi.com
 */

// used for communication via xbee api
import processing.serial.*;

// xbee api libraries available at http://code.google.com/p/xbee-api/
// Download the zip file, extract it, and copy the xbee-api jar file
// and the log4j.jar file (located in the lib folder) inside a "code"
// folder under this Processing sketch's folder (save this sketch, then
// click the Sketch menu and choose Show Sketch Folder).
import com.rapplogic.xbee.api.XBee;
import com.rapplogic.xbee.api.XBeeAddress64;
import com.rapplogic.xbee.api.XBeeException;
import com.rapplogic.xbee.api.XBeeTimeoutException;
import com.rapplogic.xbee.api.zigbee.ZNetRemoteAtRequest;
import com.rapplogic.xbee.api.zigbee.ZNetRemoteAtResponse;

import com.rapplogic.xbee.api.ApiId;
import com.rapplogic.xbee.api.AtCommand;
import com.rapplogic.xbee.api.AtCommandResponse;
import com.rapplogic.xbee.api.XBeeResponse;
```

```
import com.rapplogic.xbee.api.zigbee.NodeDiscover;

String version = "1.02";

// *** REPLACE WITH THE SERIAL PORT (COM PORT) FOR YOUR LOCAL XBEE ***
String mySerialPort = "/dev/tty.usbserial-A1000iMG";

// create and initialize a new xbee object
XBee xbee = new XBee();

int error=0;

// make an array list of switch objects for display
ArrayList switches = new ArrayList();
ArrayList nodes = new ArrayList();

// create a font for display
PFont font;
float lastNodeDiscovery;


void setup() {

  size(800, 230); // screen size
  smooth(); // anti-aliasing for graphic display

  // You'll need to generate a font before you can run this sketch.
  // Click the Tools menu and choose Create Font. Click Sans Serif,
  // choose a size of 10, and click OK.
  font = loadFont("SansSerif-10.vlw");
  textFont(font);

  // The log4j.properties file is required by the xbee api library, and
  // needs to be in your data folder. You can find this file in the xbee
  // api library you downloaded earlier
  PropertyConfigurator.configure(dataPath("")+"log4j.properties");

  // Print a list in case the selected serial port doesn't work out
  println("Available serial ports:");
  println(Serial.list());
  try {
    // opens your serial port defined above, at 9600 baud
    xbee.open(mySerialPort, 9600);
  }
  catch (XBeeException e) {
    println("");
    println("  ** Error opening XBee port: " + e + " **");
    println("");
    println("Is your XBee plugged in to your computer?");
    println("Did you set your COM port in the code near line 30?");
    error=1;
  }

  // run a node discovery to find all the radios currently on the network
  // (this assumes that all the network radios are Actuator nodes)
```

```
    nodeDiscovery();
    lastNodeDiscovery = millis(); // note the time when the discovery was made
}


// draw loop executes continuously
void draw() {

  background(255); // draw a white background

  // report any serial port problems in the main window
  if (error == 1) {
    fill(0);
    text("** Error opening XBee port: **\n"+
      "Is your XBee plugged in to your computer?\n" +
      "Did you set your COM port in the code near line 27?",
      width/3, height/2);
  }

  // create a switch object for each node that doesn't have one yet
  // ...and get current state of every new node
  for (int j=0; j < nodes.size(); j++) {
    XBeeAddress64 address64 = ((NodeDiscover) nodes.get(j)).getNodeAddress64();
    int i = 0;
    boolean foundIt = false;
    for (i=0; i < switches.size(); i++) {
      if ( ((Switch) switches.get(i)).addr64.equals(address64) ) {
        foundIt = true;
        break;
      }
    }

    // if the switch does not yet exist, create a new one
    // stop if there's more than can fit on the screen
    if (foundIt == false && switches.size() < 5) {
      switches.add(new Switch(address64, (switches.size())));
      ((Switch) switches.get(i)).getState();
    }
  }


  // draw the switches on the screen
  for (int i =0; i<switches.size(); i++) {
    ((Switch) switches.get(i)).render();
  }


  // periodic node rediscovery
  if (millis() - lastNodeDiscovery > 15 * 60 * 1000) { // every 15 minutes
    nodeDiscovery();
    lastNodeDiscovery = millis();
  }
} // end of draw loop
```

```
// function to look up all the nodes on the network
// and add them to an ArrayList
void nodeDiscovery() {

  long nodeDiscoveryTimeout = 6000;
  nodes.clear(); // reset node list, removing all old records
  switches.clear(); // reset switch list, removing all old records
  print ("cleared node list, looking up nodes...");

  try {
    println("sending node discover command");

    // send the node discover command:
    xbee.sendAsynchronous(new AtCommand("ND"));
    long startTime = millis();

    // spend some time waiting for replies:
    while (millis() - startTime < nodeDiscoveryTimeout) {
      try {
        // look for incoming responses:
        XBeeResponse response = (XBeeResponse) xbee.getResponse(1000);

        // check to make sure it's a response to an AT command
        if ( response.getApiId() == ApiId.AT_RESPONSE) {
          // parse the node information from the response:
          NodeDiscover node = NodeDiscover.parse((AtCommandResponse)response);
          nodes.add(node); // add the node to an existing Array List
          println("node discover response is: " + node);
        }
        else {
          // println("ignoring unexpected response: " + response);
        }
      }
      catch (XBeeTimeoutException e) {
        print("."); // prints dots while radio lookups are in progress
      }
    }
  }
  // if the ND response times out, note the error
  catch (XBeeTimeoutException e) {
    println("request timed out. make sure your " +
            "remote XBee is configured and powered on");
  }
  // if some other error happens, print it to the status window
  catch (Exception e) {
    println("unexpected error" + e);
  }
  println("Node Discovery Complete");
  println("number of nodes: " + nodes.size());
}
```

```
// this function runs once every time the mouse is pressed
void mousePressed() {
  // check every switch object on the screen to see
  // if the mouse press was within its borders
  // and toggle the state if it was (turn it on or off)
  for (int i=0; i < switches.size(); i++) {
    ((Switch) switches.get(i)).toggleState();
  }
}



// defines the switch objects and their behaviors
class Switch {

  int switchNumber, posX, posY;
  boolean state = false; // current switch state
  XBeeAddress64 addr64;  // stores the raw address locally
  String address;        // stores the formatted address locally
  PImage on, off;        // stores the pictures of the on and off switches


  // initialize switch object:
  Switch(XBeeAddress64 _addr64, int _switchNumber) {
    on = loadImage("on.jpg");
    off = loadImage("off.jpg");
    addr64 = _addr64;
    switchNumber = _switchNumber;
    posX = switchNumber * (on.width+ 40) + 40;
    posY = 50;

    // parse the address int array into a formatted string
    String[] hexAddress = new String[addr64.getAddress().length];
    for (int i=0; i<addr64.getAddress().length;i++) {
      // format each address byte with leading zeros:
      hexAddress[i] = String.format("%02x", addr64.getAddress()[i]);
    }
    // join the array together with colons for readability:
    address = join(hexAddress, ":");

    println("Sender address: " + address);
  }

  void render() { // draw switch on screen
    noStroke(); // remove shape edges
    if(state) image(on, posX, posY); // if the switch is on, draw the on image
    else image(off, posX, posY);     // otherwise, if the switch is off,
                                     // draw the off image
    // show text
    textAlign(CENTER);
    fill(0);
    textSize(10);
    // show actuator address
    text(address, posX+on.width/2, posY + on.height + 10);
    // show on/off state
```

```
    String stateText = "OFF";
    fill (255,0,0);
    if (state) {
      stateText = "ON";
      fill(0,127,0);
    }
    text(stateText, posX + on.width/2, posY-8);
}

// checks the remote actuator node to see if it's on or off currently
void getState() {
  try {
    println("node to query: " + addr64);

    // query actuator device (pin 20) DO (Digital output high = 5, low = 4)
    // ask for the state of the DO pin:
    ZNetRemoteAtRequest request= new ZNetRemoteAtRequest(addr64, "DO");

    // parse the response with a 10s timeout
    ZNetRemoteAtResponse response = (ZNetRemoteAtResponse)
      xbee.sendSynchronous(request, 10000);

    if (response.isOk()) {

      // get the state of the actuator from the response
      int[] responseArray = response.getValue();
      int responseInt = (int) (responseArray[0]);

      // if the response is good then store the on/off state:
      if(responseInt == 4|| responseInt == 5) {
        // state of pin is 4 for off and 5 for on:
        state = boolean( responseInt - 4);
        println("successfully got state " + state + " for pin 20 (DO)");
      }
      else {
        // if the current state is unsupported (like an analog input),
        // then print an error to the console
        println("unsupported setting " + responseInt + " on pin 20 (DO)");
      }
    }
    // if there's an error in the response, print that to the
    // console and throw an exception
    else {
      throw new RuntimeException("failed to get state for pin 20. " +
                                 " status is " + response.getStatus());
    }
  }
  // print an error if there's a timeout waiting for the response
  catch (XBeeTimeoutException e) {
    println("XBee request timed out. Check remote's configuration, " +
            " range and power");
  }
  // print an error message for any other errors that occur
  catch (Exception e) {
    println("unexpected error: " + e + "  Error text: " + e.getMessage());
```

```
        }
    }

    // this function is called to check for a mouse click
    // on the switch object, and toggle the switch accordingly.
    // it is called by the MousePressed() function so we already
    // know that the user just clicked the mouse somewhere
    void toggleState() {

        // check to see if the user clicked the mouse on this particular switch
        if(mouseX >=posX && mouseY >= posY &&
            mouseX <=posX+on.width && mouseY <= posY+on.height)
        {
            println("clicked on " + address);
            state = !state; // change the state of the switch if it was clicked

            try {
                // turn the actuator on or off (pin 20)
                // DO (Digital output high = 5, low = 4)
                int[] command = {
                    4
                }; // start with the off command
                if (state) command[0]=5; // change to the on command
                                         // if the current state is on
                else command[0]=4; // otherwise set the state to off

                ZNetRemoteAtRequest request =
                    new ZNetRemoteAtRequest(addr64, "DO", command);
                ZNetRemoteAtResponse response =
                    (ZNetRemoteAtResponse) xbee.sendSynchronous(request, 10000);

                // if everything worked, print a message to the console
                if (response.isOk()) {
                    println("toggled pin 20 (DO) on node " + address);
                }
                // if there was a problem, throw an exception
                else {
                    throw new RuntimeException(
                        "failed to toggle pin 20.  status is " + response.getStatus());
                }
            }
            // if the request timed out, print
            // that error to the console and
            // change the state back to what
            // it was originally
            catch (XBeeTimeoutException e) {
                println("XBee request timed out. Check remote's " +
                        "configuaration, range and power");
                state = !state;
            }
            // if some other error occured, print that
            // to the console and change the state back
            // to what it was originally
            catch (Exception e) {
                println("unexpected error: " + e +
```

```
                    "  Error text: " + e.getMessage());
            state = !state;
        }
      }
    }
} // end of switch class
```

Summary

One minute you're collecting meaningful data from a sensor network, and the next you're remotely activating your home appliances. It is not a stretch to consider yourself a wireless networking expert at this point, easily traversing multiple nodes in a single bound! Now you may be wondering how to make the next big leap. How do you push your powers beyond the surly bonds of ZigBee to touch other networks, like the Internet? The next chapter will unfurl the glory of gateways, opening up a pathway for your wireless networks to talk to almost anything or anyone, in whatever protocol is spoken in that realm. Take a moment to pat yourself on the back first. You deserve it.

CHAPTER 7

Over the Borders

ZigBee is only one of the great flavors of networking. In this chapter we learn to make gateways that cross borders to connect with neighboring networks, including a remarkably easy path to the Internet. You'll see full examples, showing how to allow anything to talk to everything everywhere; plus there's something special for you starry-eyed celebrity fans. Let this chapter be your passport.

Gateways

The great thing about standards is there are so many of them. Bluetooth, IPv6, UDP, ZigBee, SMS, VoIP, WiFi, Ethernet, 4G, SMTP, and TCP/IP all define different networking protocols and layers within those protocols. There's no such thing as a perfect network; that's why there are so many different ways to get networking done. Each protocol is designed to solve a specific type of problem. Most do a great job at their task. For example, Bluetooth performs solidly when connecting up eight local devices as a personal area network. At the same time, many of the engineering choices that solve one kind of problem create barriers when the protocol is pushed into unfamiliar territory. Bluetooth's simple pairing and addressing schemes don't readily scale to networks of hundreds or thousands of nodes, while other protocols may sacrifice such simplicity for flexibility and scalability. Luckily, there's no need to stick to a single protocol. Each can do what it's best at, and connections can be made so they all work well together. Using the best network means using many networks. Gateways are the glue that holds them all together.

A gateway is any device that provides connectivity between different networks. In some cases, the two networks use the same protocol and are separated by a gateway for traffic or security reasons. These won't be our focus here. We're mostly interested in gateways that expand our capabilities by opening up a world beyond the local mesh. These gateways will allow the information we've made available with ZigBee to traverse a whole web of interesting networks, accomplishing worthwhile and sometimes extraordinary tasks.

XBee as Embedded Gateway

Any XBee radio that is using local serial communications is actually acting as a gateway between two very important protocols. As you are well aware of by now, the XBees use ZigBee to communicate wirelessly between radios. They simultaneously use TTL (sometimes called board-level or logic-level) serial to communicate over metal wiring to other local devices, such as microcontrollers and desktop computers. Everything that happens on the RX and TX pins of the XBee is using TTL serial. Everything that happens over our XBee's radio antenna is ZigBee. It's the XBee's internal circuitry and software that serves as a translator between these two protocols. This is how your computer or Arduino, which only speak to the XBee using serial over wires, gain the capability to talk using radio waves to remote devices. The XBee acts as their gateway and extends them onto ZigBee networks. That's pretty important, and it's only the tip of the iceberg when it comes to the power of gateways.

Other Embedded Gateways

Many other gateway modules are available to connect from the TTL serial communications commonly found on circuit boards to a myriad of other useful protocols. In many cases, simply wiring TX/RX on the XBee directly to RX/TX on the other embedded module effectively creates a bare-bones gateway between the two protocols, as long as the other device has been properly configured for transparent retransmissions. Here are just a few of the embedded possibilities you could explore:

Bluetooth

> The Bluetooth protocol is commonly used for small, short-range mobile personal area networks. Roving Networks makes a variety of embedded modules, including the RN-41, available on a breakout board for prototyping from SparkFun (SFE WRL-00582). This module could be used to link your ZigBee network to Bluetooth Serial Port Profile to communicate directly with certain mobile phones. (See *http: //www.sparkfun.com/*.)

CAN (or CAN-bus)

> Controller-Area Networking is a standard form of communication used widely in the automotive industry for moving information between the various devices and subsystems inside cars and trucks. Every time you take a drive, your brakes, engine, airbags, and transmission are probably communicating with CAN. The Microchip MCP2515 CAN controller and MCP2551 CAN transceiver are available on an Arduino shield that you could use to stream data to and from your car. (See *http:// www.skpang.co.uk/catalog/arduino-canbus-shield-with-usd-card-holder-p-706 .html*.)

Ethernet

> This is a big category. Ethernet is the primary wired interface to the Internet, and embedded modules are just one of many ways to bridge our communications worldwide. One useful embedded gateway is the Lantronix XPort, which can

transparently connect TTL serial signals (RX/TX) to Ethernet and TCP/IP, thereby forging a circuit-board-level connection to the Internet (module: *http:// www.lantronix.com/device-networking/embedded-device-servers/xport.html*, breakout board: SFE BOB-08845). Later in this chapter we will examine some external Internet gateways, including a powerful system that makes device communications almost as simple as using a web browser.

GPRS, 3G, 4G cellular modems

Getting data out into the world isn't all about wires. As long as your project is not too far away from the places frequented by humans, you are probably within range of the mobile wireless network. The Telit GE865, for example, is an embedded module that can bridge from your XBee's TX/RX pins to GPRS mobile data networks and all the way out to the Internet. It can also provide additional onboard logic using its built-in Python interpreter. Since the GE865 doesn't have pins that connect to a breadboard, you'll almost certainly want to start with an evaluation system (SFE CEL-09342) that makes the module ready for prototyping.

HomePlug

The HomePlug Alliance specifies a protocol for creating networks over residential electrical wiring. Though just one of many powerline networking protocols, it has gained traction recently as a popular profile for managing audio/visual systems. There is a HomePlug modem board available (SFE SEN-09080) to gateway TTL serial to HomePlug, and it even includes a pin-compatible XBee header so you can plug a radio right into it! Also see X-10 below for an older method of accomplishing communications over power lines.

RF without protocols

Certain very inexpensive transmitter modules (SFE WRL-08946) can be paired with receiver modules (SFE WRL-08950) for a low-cost one-way link. In general the reliability and flexibility of these connections is poor enough that they can't be recommended. However, their limitations often create a fine demonstration of how much ZigBee's addressing, network infrastructure, and error handling are helping you out. Proceed with caution, but don't be afraid to try them.

RFID

Radio Frequency ID tags (RFID) are small microchips that typically use radio energy as a passive power source for transmitting a serial number. RFID readers (for example, SFE SEN-08419) can detect these transmissions and repeat them as a TTL serial TX that could be retransmitted through your XBee radio. RFID is a lot like a bar code and sometimes disappoints those who try to extend it beyond its limits.

USB

Computers commonly communicate with other devices using the Universal Serial Bus protocol. Several different USB-to-TTL serial gateways have already been described in Chapter 1 since they are required for serial configuration of the XBee. Many of these use a very popular microchip from FTDI so computer drivers are readily available (*http://www.ftdichip.com/FTDrivers.htm*). This FTDI chip is available on breakout boards (SFE BOB-00718), making it easy to create generic connections between board-level TTL serial and USB host devices like computers.

WiFi

Another very popular on-ramp to the Internet is WiFi, the wireless networking protocol generally used by laptop computers and many smartphones. There are so many modules to keep track of in this space that an entire book could be written about them. Two worth knowing about are the Lantronix MatchPort (*http://lantronix.com*) because its configuration matches the Ethernet XPort, and the WiFly module because it's available on a breakout board (SFE WRL-09333) for prototyping. Remember that WiFi connections need to be configured with a different system name and security key each time they attach to a new network, so unless your device sits in one place, or incorporates a screen and keyboard, WiFi might not work for your project. Think this through before getting started.

X-10

One of the oldest home automation protocols is X-10. Originally devised as a powerline networking protocol, it has since been extended to wireless as well. Newer protocols like HomePlug may someday render X-10 obsolete but today it remains in wide use so you might have cause to use a gateway module like the TW523 (*http://www.x10.com/products/x10_tw523.htm*) to control existing home automation systems.

Z-Wave

The Z-Wave Alliance defined this proprietary wireless communications protocol, which has gained some traction in the home automation market. Like ZigBee, it is designed for low-power, low-bandwidth data interactions. It operates in a different frequency spectrum (900 MHz versus ZigBee's 2.4 GHz) and, unlike ZigBee, the protocol itself is only available under a nondisclosure agreement. Development kits (DK 703-1056-ND) are probably the best way to start working with Z-Wave, though they are far more expensive than the modules themselves (DK 703-1023-ND).

Internet Gateways

Of all the places to take your data, nothing is quite as compelling as passing it through a gateway to the Internet. That's because the Internet reaches almost everywhere and has the capability to touch almost anything. Some people think of the Internet as mostly web pages, but that's only one of a dazzling array of destinations for the data streaming from your device or sensor network. Really the Internet is a vast collection of pathways and services that has already grown so complex that it is sometimes described as beyond the grasp of human comprehension. Luckily you don't need to understand the whole thing to move information from place to place in a reasonably efficient manner. There are a lot of reasons to do it:

Data storage

Any sensor network that is devoted to amassing data will need to store that data someplace. It may be fine to bring the data onto a local computer and work with it there, but in many cases that isn't going to be enough. If there's a lot of data it may be more than you can store locally. Also there may be other people who are interested in access to your data, such as your project partners, colleagues, clients, customers, or even the general public. The great thing about the Internet is that storage can really be anywhere. It's quite common to stash data in places whose physical location is totally unknown to the user. Amazon's S3 service, for example, synchronizes your data onto a suite of servers located around the world. Customers of the service do not generally know the actual locations and physical storage media that hold their data. They usually don't care because access is far more important than the mechanics involved. Whether your data resides on a private MySQL database with a hosting provider such as Dreamhost (*http://dreamhost.com*) or HostMonster (*http://hostmonster.com*), or in a shared compendium of public data like Pachube's system (*http://pachube.com*), the advantages of outsourcing your data storage will be the same. Storage can be endlessly expanded and access can occur from any place at any time.

Data presentation

Everyone can see the Internet, so it's a good place to show your data. You may decide to roll your own display using something like Processing, or hook up with any of the online data visualization services like Google Charts (*http://code.google .com/apis/charttools/*) or Microsoft's Pivot Viewer. Data sharing sites like Pachube (see Chapter 8) often embed other sites' visualization services, so you may be using these big companies' systems indirectly. Once your data is on the Internet, there's no end to the ways it can be displayed to tell its story.

Remote actuation

Objects are on the Internet, and some of them move. Robots, door locks, plant watering systems, lights, bells, whistles, and interactive kinetic sculptures all can take data from remote sensors to perform actions based upon real-world information. Remember that a sensor can be as simple as a switch. Perhaps you want to make a lamp in London react to the current amount of daylight in Delhi. Or maybe you want the tide in Tiburon to create motion in Melbourne. Unless you're going to lay the cable yourself, the Internet is the best way to join your widely separated inputs and outputs together to create something amazing.

Everything

There's no reason to do just one of these things. You will have the most fun doing them all at the same time, uploading your data to a system that creates remote actuations (perhaps generating sensor data of its own at the other end), and creating an interface to control the systems and present their resulting tale. The Internet is big and it is new, so it presents endless opportunities to do things that have never been done before. Make your mark.

Internet Media

The why of Internet gateways should be clear and compelling. Let's take a look at the how. In most cases you'll be choosing among three options for your physical network connection: wired Ethernet, local WiFi connections, or mobile data connections that use the cell phone network. Of course there are other possibilities, like old-school telephone modems or amateur packet radio that might apply to special situations, but those won't be discussed here:

Ethernet
> Ethernet uses physical wiring, which means your gateway will be tethered in place. Ethernet connections are fast and very reliable because their wires are not exposed to much noise disruption. Configuration is often not necessary at all. Ethernet is also very cheap to physically implement so generally this will be the lowest-cost option.

WiFi
> WiFi provides a high-bandwidth wireless connection that's locally available in many homes, schools, businesses, and even some public parks. Configuration is almost always required to select the network you want to attach to and to issue the password or keys that secure each system. WiFi connections (wireless using IEEE 802.11x standards) typically communicate using unlicensed parts of the radio spectrum, which means they need to be tolerant of noise, and their physical components are more expensive. Generally, using a WiFi gateway will cost more and be somewhat less reliable than using Ethernet; however, in certain locations it may be the only practical choice.

Mobile data
> This type of connection goes by various names, including carrier wireless, cellular data, mobile Internet, GPRS, 2G/3G/4G, and several other vague or inscrutable monikers. Data plans are offered by large carriers such as Orange, AT&T, Verizon, or NTT. Mobile data is available at most locations that have mobile phone coverage. The components are similar in cost to WiFi, but the connection itself is generally far more expensive. An account with a carrier is required and needs to be provisioned in advance. Data charges can quickly accumulate into an exorbitant bill. Even so, mobile data reaches to places where Ethernet and WiFi are not a possibility. Because configuration is not location-specific, this is the best choice for any gateway that will be moving from place to place. If you want to wire up a freight truck with sensors, mobile data services will almost certainly be part of your solution.

Computers Versus Dedicated Devices

Most Internet gateways are either implemented on a personal computer or manufactured as dedicated devices. It is certainly possible to build your own dedicated gateway from scratch, though in most cases it will cost more and do less than one of the commercially available ones.

Computers are a good choice for gateways in quick prototype systems. Most likely you already have a computer, and that computer almost certainly comes with all the equipment needed to connect to both Ethernet and WiFi for access to the Internet. It's a safe bet that it also has a serial port (probably USB) for plugging in your XBee radio. In a certain sense using your computer is free, assuming you already own one. It also contains an extremely powerful processor that will have no problem performing complex manipulations to the information passing through it. So for a quick prototype, you can write some code in any number of languages (including Processing) to translate your data from ZigBee to TCP/IP.

Many people find that personal computers don't work as well for projects installed over the long term. For one thing, computers run a very complex operating system that needs regular upgrades to keep it stable and secure. In the course of running other programs, you may slow down or crash the system, disabling your gateway at the same time. Your computer also uses quite a bit of electricity, takes up a fair amount of physical space, and in the case of a personal laptop, periodically gets moved away from the location where the gateway operates. This is where a dedicated device can do a better job.

Dedicated gateways are simple pleasures. They typically come in the form of a nondescript rectangular box. For ZigBee gateways there's a radio to talk to your wireless sensor network, often a small processor chip, and another module to talk to whatever you're gatewaying to, typically either an Ethernet module, WiFi radio, or mobile data system. These types of gateway devices rarely have any kind of screen or keyboard. Configuration is usually done through a web browser or another type of remote connection. They also tend to be small, use far less power than a full-size computer, and run a slimmed-down operating system that has only a few essential features. Finally, they tend to cost less than a full computer and run stably without rebooting, potentially for years at a time. Some of them are even designed to use renewable energy sources, operate outdoors, or survive in harsh environments with extended temperature ranges. There are many flavors and brands of gateway. Since we're already working with XBee radios, we'll examine the ConnectPort line of devices made by Digi International. These are specifically designed to work seamlessly with all the features that the XBee has to offer. Other ZigBee gateways are available from Pervasa, RFM, Crestron, Exegin, AMX, Alektrona, and many others.

ConnectPorts

Digi's ConnectPort line of gateways provides many advantages: they use the XBee radios you're already familiar with, and they have an easy-to-use web interface and an internal Python programming language interpreter. Having Python inside means you can write and run programs to manipulate your incoming and outgoing data. You can provide your sensor readings in just the way a remote system wants them, interpret remote commands to turn them into meaningful actions, or do a little of both at the same time so that your sensors and actuators can work locally together, only contacting the outside world when it's essential to do so. The internal Python interpreter comes preloaded with XBee libraries that make it very easy to script commands, communications, and transformations right inside the gateway.

Selecting a ConnectPort

All of the ConnectPort models described below (and shown in Figure 7-1) support connections between ZigBee and another Internet medium. Many are optionally offered with 802.15.4 Series 1 radios (not compatible with ZigBee), so be sure you choose the ZigBee version when you purchase your ConnectPort! Some support serial/USB connections to control devices that are plugged into the gateway itself. Here we'll focus on the Internet connectivity because that's what's important for our networks:

ConnectPort X8
> This gateway supports ZigBee, Ethernet, WiFi, two different mobile data networks, two USB ports, one standard serial port, and a local sensor port. This Cadillac of the ConnectPort line costs from $800 to $1,000 depending upon configuration.

ConnectPort X5
> Designed for Vehicle Area Networks (winkingly abbreviated VAN), this gateway comes with ZigBee, satellite radio, mobile data, WiFi, and a GPS feed. It is a rugged unit designed with fleet trucks in mind. The X5 runs about $1,000.

ConnectPort X4
> The X4 router is available with Ethernet or WiFi and a slot for a mobile data radio along with one USB port and one plain serial port. A ConnectPort X4 runs from $450–700 depending upon configuration.

ConnectPort X3
> This is a brand-new and fairly inexpensive option for mobile data connections from ZigBee. The X3 offers a GSM/GPRS radio that with the addition of a data plan will let you connect from just about anywhere. The base cost is about $250.

ConnectPort X2
> Here's the simplest option. The base X2 comes with a ZigBee radio and one Ethernet port. (There's also a WiFi version that costs more). The X2 isn't brimming with memory but you can get a surprising amount done inside its 8 MB of RAM. The X2 was originally manufactured with a metal enclosure and an external antenna that retailed for about $200. Recently a slimmed-down design has been released at $99, with a plastic case that allows for an internal antenna (X2-Z11-EC-A). That's the one to get started with.

Figure 7-1. The ConnectPort line of gateways, including, from left to right: the X8 with a wide offering of interfaces, the inexpensive X2 for Ethernet, and the midrange X4, which can also come configured for WiFi or mobile data

Setting Up a ConnectPort

It is extremely easy to set up a ConnectPort. This section will demonstrate setup and configuration for the low-cost ConnectPort X2 ZB Ethernet (Digi X2-Z11-EC-A). Setting up the X4 or X8 is very similar:

1. Begin by plugging the ConnectPort into a wall outlet using its supplied power adapter.

2. Next, use a standard Ethernet cable to attach the ConnectPort to any available Ethernet jack on your Internet router. Your home network is probably already configured to assign IP addresses using DHCP, in which case simply powering it on and plugging it into the Ethernet port will allow it to configure its own network settings. If not, you'll have an opportunity to do manual configuration in the next step.

3. Connect a computer running Windows to the same network as the ConnectPort. In many cases, using the local WiFi connection will be fine, but if you aren't sure, plugging into the same Ethernet switch as your ConnectPort will ensure that you're using the same local network. Business and educational networks often have additional restrictions that your system administrator will need to help you with.

4. Some ConnectPorts come with a software CD that includes a Windows program called *Device_Discovery.exe*. Insert the CD into your Windows computer and launch Gateway Software→ConnectPort X→Configuration→Device Discovery. This will open a window that will show you all the devices on your local network. (If you don't have the CD, the program can also be downloaded from Digi's website at *http://www.digi.com/support/getasset.jsp?fn=40002265&tp=4*.)

Troubleshooting

If you don't see any devices listed in the Device Discovery program, check to make sure that the ConnectPort is showing a power light on the front, and that it is properly connected to Ethernet. A yellow link light on the Ethernet jack will illuminate as long as there is some kind of Ethernet connection active. Also make sure your computer is on the same network. Some home Internet setups are multiple boxes that operate on different TCP/IP networks from each other. If possible, try plugging your computer into the same set of Ethernet jacks as your ConnectPort, turn off WiFi, and see if clicking on "Refresh view" convinces the ConnectPort to pop up in your device list, as shown in Figure 7-2.

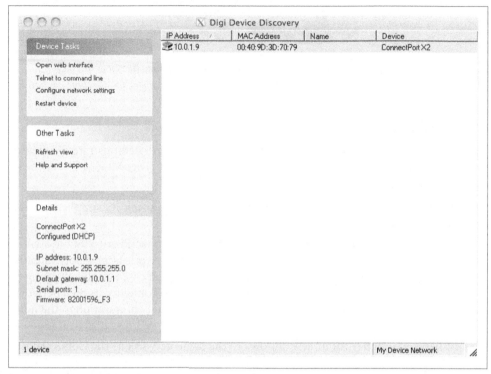

Figure 7-2. Using Device_Discovery.exe to locate the ConnectPort on your local TCP/IP network

 If your network does not assign IP addresses automatically (via DHCP), click on the listed device to select it, then click "Configure network settings" under Device Tasks to enter the IP address information for your network.

Configuring a ConnectPort

Once you can see the ConnectPort listed in the Device Discovery window, make a note of its IP address. In most cases double-clicking on the device in the list will open a web browser; however, if it doesn't, you can also simply type its IP address into your browser's URL field. For example, if the IP address listed is 10.0.1.9, putting *http://10.0.1.9* into your web browser should open up a configuration screen similar to the one shown in Figure 7-3.

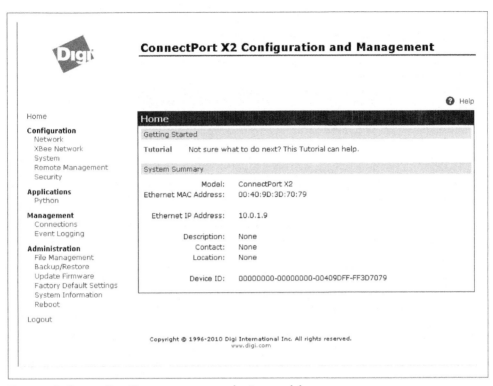

ConnectPort X2 Configuration and Management

                                                                    ❓ Help

Home

Configuration
  Network
  XBee Network
  System
  Remote Management
  Security

Applications
  Python

Management
  Connections
  Event Logging

Administration
  File Management
  Backup/Restore
  Update Firmware
  Factory Default Settings
  System Information
  Reboot

Logout

Home

Getting Started

Tutorial   Not sure what to do next? This Tutorial can help.

System Summary

                       Model: ConnectPort X2
       Ethernet MAC Address: 00:40:9D:3D:70:79

       Ethernet IP Address: 10.0.1.9

              Description: None
                  Contact: None
                 Location: None

              Device ID: 00000000-00000000-00409DFF-FF3D7079

Figure 7-3. ConnectPort Home screen accessed using a web browser

 If you are familiar with the Telnet program, you can use it to contact the ConnectPort by connecting to its IP address on the default port 23 for a command-line interface. Type a ? at the prompt to get a list of valid commands. Then type a ? after any command name to examine available options.

The ConnectPort comes preconfigured to obtain its IP address automatically via DHCP. If you'd prefer that it had a fixed address, or if you need to change any other TCP/IP network settings, they can be accessed by clicking on Network, displaying the screen shown in Figure 7-4.

Inside the ConnectPort is a ZigBee radio that is preconfigured to be a network coordinator. To view or change its settings, first click on XBee Network to see a list like the one shown in Figure 7-5. This list will include the internal radio, as well as any other devices that have joined the ZigBee network.

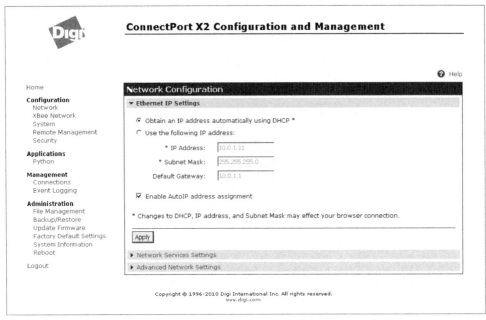

Figure 7-4. Network configuration for the ConnectPort's TCP/IP connection to the Internet

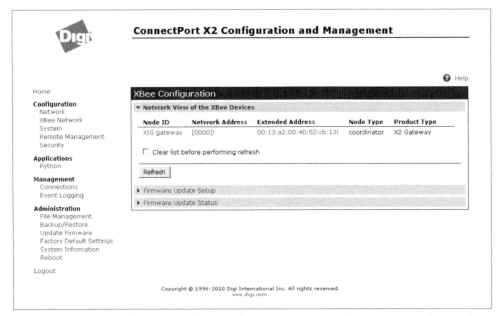

Figure 7-5. XBee Network lists the radio inside the ConnectPort, as well as any other radios that have joined the network

Each radio listed in the XBee Network screen can be configured through your browser. This includes the local radio inside the ConnectPort. But wait, there's more. It also includes any other remote radio that has joined the ConnectPort's network. You can now change the configurations of all the radios on your network from the comfort of your laptop computer by clicking on that XBee in the list to see its detail screen, as shown in Figure 7-6. These can be changed right from your browser! You may notice that these figures show a node ID that labels the radio. See the sidebar "Naming Radios" on page 202 for information about how to set and discover these node identifiers.

 Be careful changing the settings on remote radios. If you make a change that accidentally causes one to leave the network, you may then need to physically access that radio to fix the situation, which might be difficult if your remote radio is miles away, or on the ceiling, or strapped to an angry goat. Think before you click!

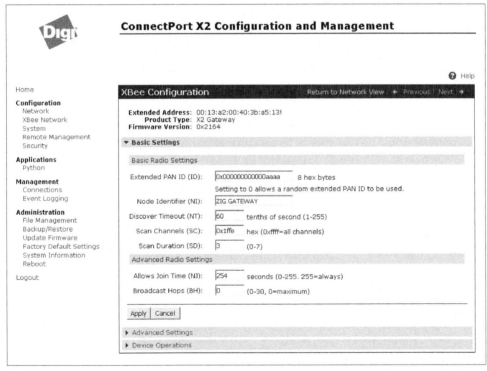

Figure 7-6. XBee Configuration details show both basic and advanced settings for every radio joined to your ConnectPort's network

We will look at several other configuration screens in detail when we set up the XBee Internet Gateway later in this chapter. In the meantime, you might want to browse around and familiarize yourself with the various settings. The Help link at the top right of each screen will link you to a basic explanation of the functions. Of course, you should be somewhat cautious about changing any settings you don't understand, but remember that measured bravery is the prime gateway to technical expertise.

 If you do something that colossally messes up the configuration, to the point where you can no longer access the ConnectPort, you can always return the device to its factory defaults by powering it up while holding down the recessed reset switch with a paper clip. Continue pressing the reset switch for 20–40 seconds. The Status light will begin blinking in a 1-5-1 pattern, indicating that the ConnectPort should now be back to its factory configuration.

Naming Radios

Each ZigBee radio has at least two addresses: its permanent and unique 64-bit address and its coordinator-assigned 16-bit address. Human beings have a much easier time remembering names than numbers, especially long numbers. In a welcome nod to the human race, XBee radios offer another, more humane option. Each radio can be configured with a node identifier text string that names it over the network. The `ATNI` command will set this Node Identifier with any phrase up to 20 characters long. For example, to set the name of your radio to Mary's Temperature Sensor, the command in AT mode would be `ATNI Mary's Temperature Sensor` (the space after `ATNI` is optional but makes for easier reading). Typing `ATNI` by itself and hitting Enter will cause the local radio to reply with its current node identifier. In most cases, though, you will want to use this human-friendly name to find a *remote* radio that's out on your network.

The Node Discovery command `ATND` will start this process. In AT mode (transparent/command mode), the Node Discovery command will be broadcast to the entire network and each radio that can hear it will respond with a block of information about itself, including its node identifier if that's been set. That format for each block you receive will be:

```
16-bit MY address
First half (SH) of 64-bit address
Second half (SL) of 64-bit address
Node Identifier text string (this line blank if NI has not been set)
Parent Network address (you can ignore this information)
Device Type (0=Coordinator, 1=Router, 2=End Device)
Status (can be ignored)
Profile ID (can be ignored)
Manufacturer ID (can be ignored)
```

In API mode, issuing this command (inside an AT command frame, of course) will result in a separate response frame being returned for each radio. The command data will contain response blocks similar to the one above.

It's also possible to configure any radio's *destination* address by using the `ATDN` command to tell it the destination's node identifier you'd like it to speak to. In AT command mode, `ATDN Mary's Temperature Sensor` will attempt to discover that node's numeric address over the air, and if it's found will automatically set the `DL` and `DH` registers to the appropriate 64-bit address. When the `ATDN` command is successful, it returns `OK` to let you know, then exits command mode immediately. If the `ATDN` command fails to find the radio you requested on the network, it will return `ERROR` and remain in command mode. If you are using API firmware, issuing this command inside an AT Command frame will result in a response frame that contains both the 16- and 64-bit addresses of the remote radio, along with a success or error indicator in the Command Status byte. The Destination Node command allows you to implement a whole system of radio naming and name lookups that bypass using any numeric addresses in your code. Consider this if your project involves creating many duplicate networks where each node is in a role that could be called by a specific name, no matter what its physical or assigned numeric address might happen to be.

Remote Management

You may still be glowing with excitement about your newfound ability to configure radios wirelessly from the ConnectPort's web interface. It is certainly pretty cool but it does require that you have *direct* network access to your ConnectPort, something that's not always possible once you leave the location where your ConnectPort lives. In most cases a firewall, network address translation, or other network obstacle of some kind means you'll need to be physically near the device and plugged into the same network to access the ConnectPort's web interface. But if you think that's going to limit your powers, well hang on to your hats and glasses because the ride is about to get about 10 times better. At the Remote Management screen you can configure your ConnectPort to open a special connection that links it to a central access server at Digi International called iDigi. Once that link is set up, logging on to the iDigi server's web interface from *anywhere* gives you full configuration access to all of your ConnectPorts *and every radio that's joined to any of those ConnectPorts* . This is a massively powerful feature. It's like suddenly discovering you are in command of a robot army, willing to do your worldwide bidding. Here's how to get started:

1. Click on Remote Management in the web interface to show the configuration screen. Check the box for "Enable Remote Management and Configuration using a client-initiated connection," then enter `developer.idigi.com` for the Server Address as shown in Figure 7-7. This is where you can link to a central server and command your robot army of sensor networks.

2. Check the box to "Automatically reconnect to the server after being disconnected." The default setting of 1 minute should be fine. This will ensure a persistent connection.

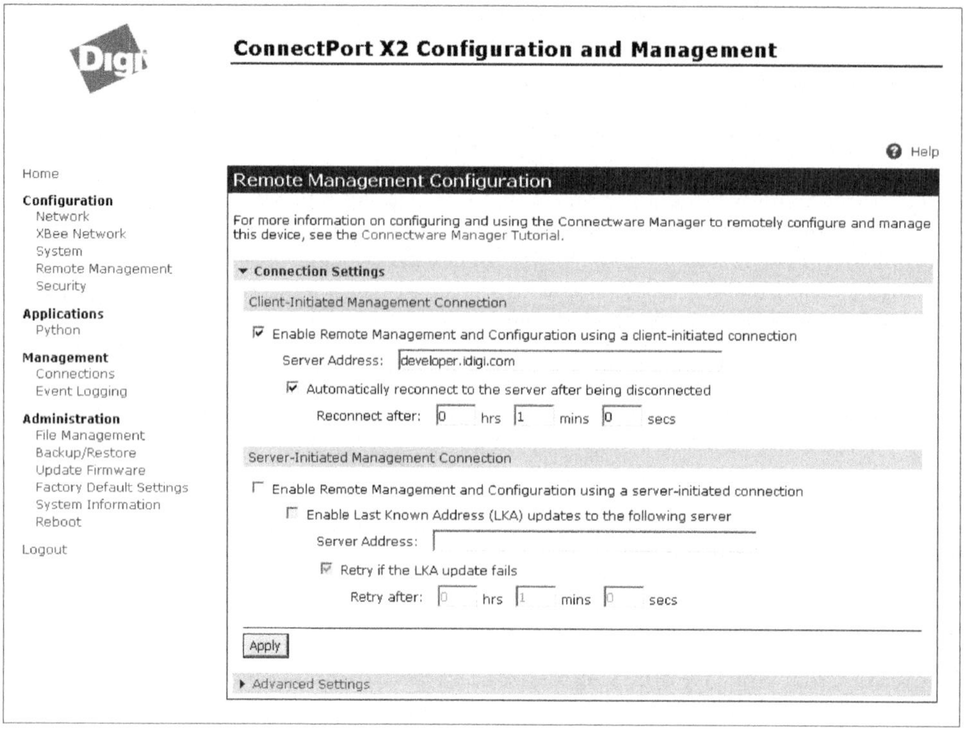

Figure 7-7. The Remote Management Configuration screen on the ConnectPort

3. Click the Apply button at the bottom of the screen to save your changes. The ConnectPort will now attempt to make an outgoing connection to the iDigi server.

iDigi Connectivity Server

iDigi is a cloud service that aggregates ConnectPorts and their networks so that they can be accessed and configured remotely from anywhere in the world. Currently anyone can set up a free account to control up to five ConnectPorts, along with an unlimited number of radios connected to each of those ConnectPorts. To begin, go to *http://developer.idigi.com* (Figure 7-8). Before you can log in, you'll need an account. Select the New User link and fill out the registration forms (Figure 7-9). At the end you'll be taken back to the login screen where you can now access your account. Enter your newly selected username and password to begin your remote networking adventure.

iDigi Features

The iDigi service is intended as a platform for application-to-device messaging, data storage, and administration. We'll mostly be considering the administration features here, because you'll want to use them with your ConnectPort.

Figure 7-8. iDigi login screen at http://developer.idigi.com

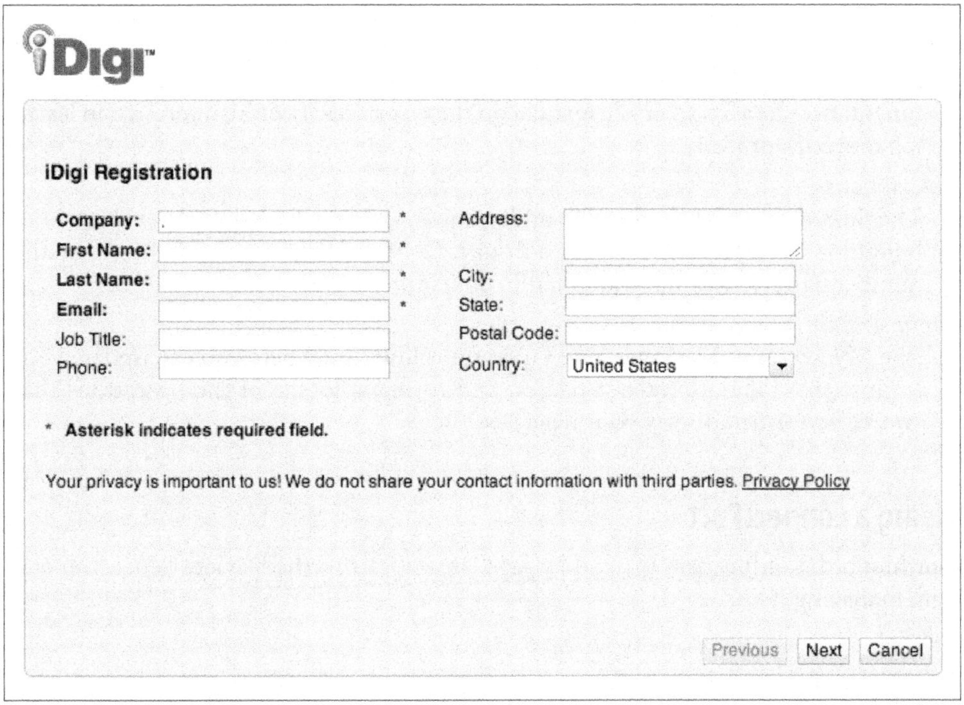

Figure 7-9. iDigi registration will set you up with a free account for remote management of up to five different ConnectPorts

On every iDigi screen (see Figure 7-10), you have access to a menu that includes:

Home

    This contains the Welcome link that describes the system and a Documentation area that connects to support forums and various downloadable documents.

Management

    Here is where most of your iDigi work will get done. It's where you can administer your ConnectPort devices and their networks of XBees, along with iDigi's data storage and messaging features.

Devices

    Every piece of equipment that connects directly to iDigi will be listed here. You'll want to add your ConnectPort to this list, and we'll show you how below.

XBee Networks

    Once some ConnectPort devices have been added, their ZigBee networks can be discovered. They will then be listed here where they can all be remotely configured.

Storage

    This is where uploaded data can be stored. Check the online documentation for more information about storage features.

Web Services Console

    Web services are a standard for exchanging information that use HTTP browser protocols and URLs as an application programming interface. This is where the application-to-device messaging features of iDigi are implemented. Check the documentation area for an entire manual on these services if you're interested in learning more about them.

Subscriptions

    The Summary and Details links under Subscriptions show you the features and limitations associated with your account. Contact iDigi if you need to raise the number of devices or allowable traffic limits for your project.

Administration

    The My Account, Messages, and Operations links are where you can update your settings, read about system updates, and review a logfile of the operations that you've performed during your iDigi session.

Adding a ConnectPort

Your first order of business is to add your ConnectPort to the Devices list so you can begin managing it.

 It is best to complete the local configuration steps described in "Remote Management" on page 203 before attempting to add your device to iDigi. This will ensure that your ConnectPort is preconfigured to open a socket connection to iDigi even if it can't be discovered automatically.

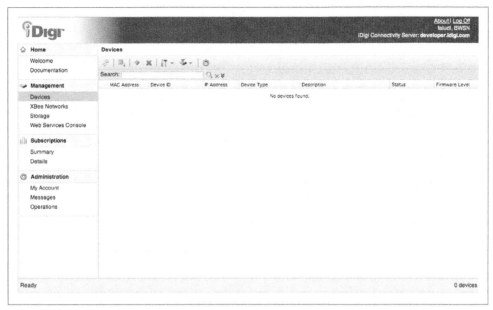

Figure 7-10. iDigi Devices screen, before any devices have been added

Select the Devices link and then click on the blue plus sign icon on the button bar to begin adding a new device. You'll see a dialog box called Add Devices, as shown in Figure 7-11.

If your ConnectPort is on the same local network as the computer where you are using iDigi's web interface, it may be listed here automatically. If it is, simply click on the device in the list and then click the OK button to add it to iDigi. Not all browsers or networks will allow this automatic discovery to happen properly, so if you don't see your device listed, try clicking the Add Manually button to see the display shown in Figure 7-12.

To add a ConnectPort manually, enter its MAC address (the unique hardware address assigned to every Ethernet device) in the MAC field and click on the Add button. You can find the MAC address printed on the back of each ConnectPort. It will begin with 00409D, which is the official prefix for all Ethernet addresses assigned to Digi devices. You can either use the XXXXXX:XXXXXX or XX:XX:XX:XX:XX:XX format. In Figure 7-12 the MAC address entered is 00:40:9D:33:B7:0C; yours, of course, will be different. Once you enter that address and click on the Add button, you can click the OK button at the bottom of the screen to complete the process of adding the new device to your iDigi account. A green bar at the top of the screen will briefly appear to indicate that your device was added successfully, as shown in Figure 7-13.

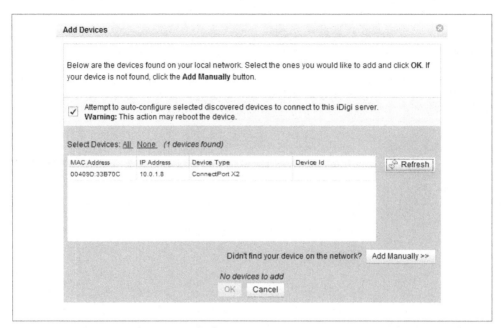

Figure 7-11. The iDigi Add Devices dialog box will automatically attempt to discover any devices that are on your local network

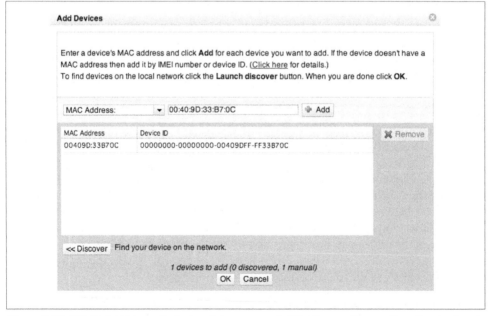

Figure 7-12. iDigi's Add Manually feature lets you search iDigi for any ConnectPort that is already configured for remote administration

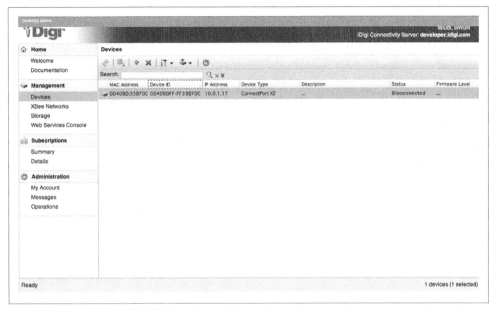

Figure 7-13. Devices newly added to iDigi will be confirmed by an announcement in green at the top of the screen

If your device's status is initially listed as Disconnected, try waiting a minute and then clicking on the circling yellow arrows in the button bar to refresh the list. Also, check your ConnectPort's connection to the Internet, and be sure that you ran through all the steps listed in "Remote Management" on page 203.

Viewing Configurations

Once you have added your ConnectPort to the Devices list and its Status is listed as Connected, simply double-click on the listed device to open up its properties page. You'll first see a Home screen as shown in Figure 7-14, along with a list of links to configure the various properties of the ConnectPort. Any system information that has been entered will be shown here as well. Now you can perform many types of configuration from anywhere in the world that you can find an Internet connection!

> Some features differ between the direct web interface and the iDigi interface. For example, iDigi does not currently show you a list of active connections or read the event log. These features may be added in the future, as iDigi is under active development.

Figure 7-14. The ConnectPort's Devices Home screen on iDigi lists its MAC address and model

For example, the Python link shows a screen (Figure 7-15) that lists all the programs and libraries loaded onto this device. iDigi gives you full access to remotely upload and delete these files, as well as to indicate which ones should start up automatically when the device is powered on. We will talk more about Python files in "XBee Internet Gateway (XIG)" on page 214.

Another useful screen is the System screen (Figure 7-16) under the Advanced Configuration link visible in Figure 7-15. Here you can enter a text description for your ConnectPort that will show up in the device list. It's very helpful to set this information if you have a number of ConnectPorts, so that it's easy to see in the listing which one is which.

Firmware Updates and Remote Reboot

Several more essential features can be accessed via the Devices list. (Clicking on the Devices tab at the top of the screen will take you back there.) Each ConnectPort can remotely receive upgrades to its internal OS firmware, as well as upgrades to the firmware that drives its *internal* XBee radio. (We'll talk about configuring the other XBees in the network below.) The Firmware icon in the button bar (Figure 7-17) displays a menu that includes these options.

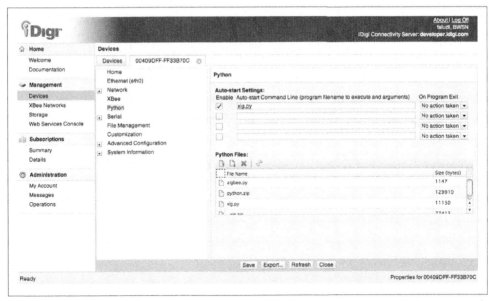

Figure 7-15. ConnectPort's Devices Python administration screen on iDigi; program files and startup can be managed here

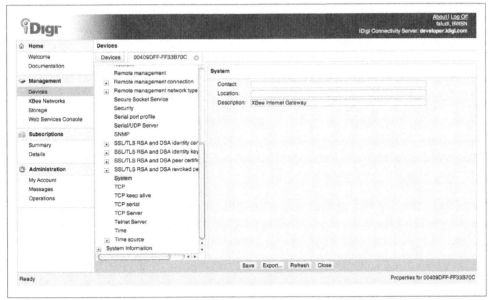

Figure 7-16. System information, including a description, can be entered under the Advanced Configuration link

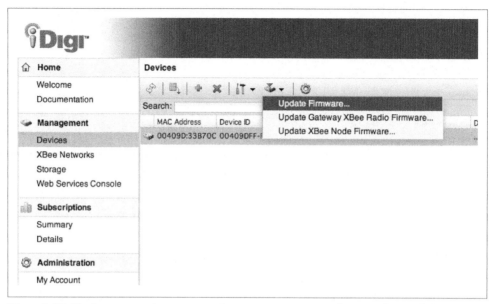

Figure 7-17. The firmware update menu on iDigi gives remote access for upgrading and changing the low-level device firmware

Firmware for the ConnectPort and the XBees can be downloaded to your local hard drive from the Support area on the regular Digi website (*http://www.digi.com/sup port*). Back at iDigi, you can select the appropriate menu item to upload firmware with the *.bin* extension for the ConnectPort and with an *.ebl* extension for the Gateway XBee. Updating the firmware on remote XBee nodes takes two steps. First, select the Update XBee Node Firmware menu item to place the appropriate *.ebl* files onto the Connect-Port. For example, to update to the version of the ZigBee Router AT that's current as of this writing, you could upload a file called *XB24-ZB_2270.ebl* that's available in the Digi website's Support area. The second step is to make sure that the ConnectPort is configured for over-the-air firmware updates. Navigate to the device's properties and select the XBee link to show a screen like the one in Figure 7-18. Check all four boxes to ensure that your update is distributed automatically over the air to any radio that's out of date. You can also upload and delete XBee firmware files using the interface on this screen. Over-the-air firmware updates are an extremely powerful feature of iDigi. As long as your radios are joined to a ConnectPort, you are able to send them new firmware over the Internet using their own radio connection. Amazing!

Viewing an XBee Network

It's easy to examine your remote XBee networks with iDigi. Click on the XBee Networks link to see a list of all the radios that have currently been discovered (Figure 7-19). Initially you'll probably see only a single radio, the XBee that's inside the ConnectPort gateway. If other radios have joined your ConnectPort's network, you can discover them by clicking on one of the flashlight icons in the button bar—the first to do a regular discovery and the second to clear the cache and rediscover the network from scratch.

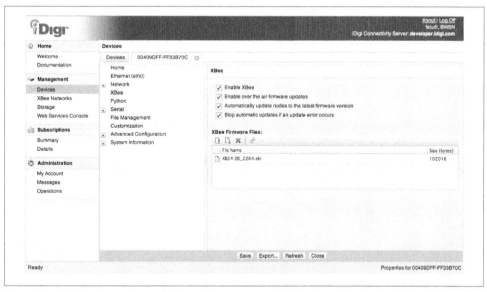

Figure 7-18. XBee firmware updates can be automatically distributed from the ConnectPort, using the configuration and files listed on the device's XBee screen

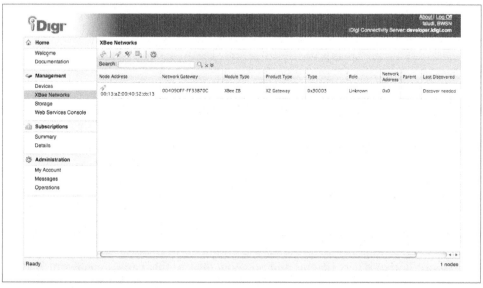

Figure 7-19. In iDigi, XBee Networks lists all the radios that have currently been discovered on all your networks, in this case just the internal gateway radio that displays initially

To configure any of the remote XBees or the gateway XBee, simply double-click on its name in the list to open up a tab with its properties, as shown in Figure 7-20. Most of the interesting settings are on the Advanced screen (Figure 7-21). Some of these are old friends because they are the exact same settings you've been configuring with AT commands all along. Now you can change those settings from anywhere! For example, you can change the node identifier by entering a new one in the field called "Node identifier" on the Basic screen and then pressing the Save button at the bottom. (See the sidebar "Naming Radios" on page 202 for more information.) Use a reasonable amount of caution when you make changes because they will be executed immediately, and any changes that cause your remote radios to leave the network might require physical access to get them to rejoin.

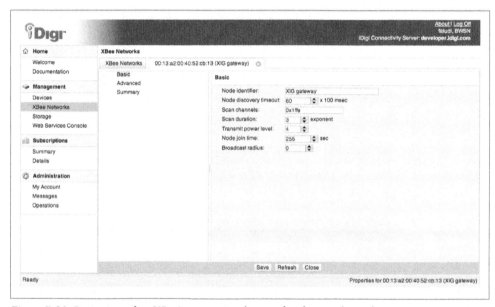

Figure 7-20. Basic view of an XBee's properties shows a few frequently used settings

Now that you've had a solid tour of the ConnectPort and iDigi's management services, you are probably eager to create a working system of your own. The next section will show you how to install and run the XBee Internet Gateway, getting you ready for the example Twitter Reader project at the end of this chapter.

XBee Internet Gateway (XIG)

The ConnectPorts are very flexible and powerful devices that can connect your ZigBee network to any Internet service in pretty much any way you like. The seemingly unlimited range of options can sometimes feel overwhelming to a beginner. Rather than learn about TCP/IP addressing, port numbering, DNS, application-layer protocols, and

Python programming—terrific as all those things are to know—you probably would like to start with something simple that opens a path between your prototype and the teeming mass of terrific services that are available on the Web.

The XBee Internet Gateway is a Python program that can be loaded onto any ConnectPort right out of the box. It's an interface that mirrors the interactions humans have in web browsers. Once XIG is running, any radio that sends it a standard plain-text URL will receive back the regular results from that URL. Redirects to other pages, timeouts, security, retries, and so forth, are all handled behind the scenes just like they are in a web browser. Take a look at View Source in your browser; you can see the web page's underlying HTML. With the XIG, the radios in your project see the exact same thing. Each can send out a URL and receive back the source for that web page—whatever it is. This simple service shifts all the hard stuff about interacting with the Web to the gateway. There's no need to handle security, domain lookups, or redirects locally. That's all taken care of for you by the XIG on the ConnectPort, giving your prototype a very simple yet completely flexible pathway to any web service you can imagine.

XIG is an open source team effort lead by your author, Jordan Husney, and Ted Hayes, with valuable support from a community of commercial and educational users. You can view the code and contribute your own efforts at *http://code.google.com/p/xig*.

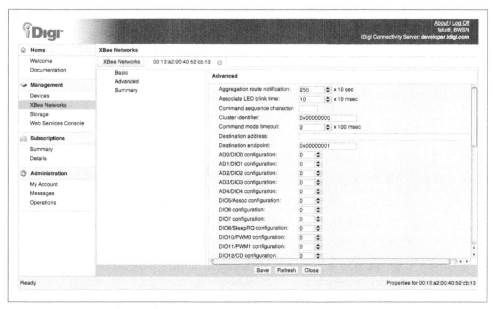

Figure 7-21. Advanced view of an XBee's properties shows many familiar AT command settings

Installing and Configuring XIG

Setting up the XIG is easy, now that you're familiar with the general administration of a ConnectPort gateway. If you haven't already, follow the instructions in the previous sections "Setting Up a ConnectPort" on page 197 and "Configuring a Connect-Port" on page 198. Next, download the XIG code, which is linked from this book's website. It should also be available at *http://code.google.com/p/xig/downloads/list* in the form of a ZIP file that contains *xig.py* and *_xig.zip*. Make sure the main file (*xig-x.x.x-bin.zip*) gets unzipped, but leave *_xig.zip* and any other internal files in their compressed form.

> The XBee Internet Gateway is still undergoing active development and may have been upgraded and changed by the time you read this. Check this book's website (see the Preface) to make sure you have the latest files and instructions.

Once you are looking at the ConnectPort's administration interface, click on the Python link to begin uploading files. The ConnectPorts all come with the required *python.zip* libraries preloaded. There may be other resource files here as well, including *zigbee.py*, which is not needed for this project but can safely remain in the directory. Click the Browse button to navigate to and select each file you'd like to upload, starting with *xig.py*. Click the Upload button and wait for a File Uploaded message to appear at the top of the screen, as shown in Figure 7-22. Repeat this process for *_xig.zip*.

Next, click the Auto-start Settings link at the bottom of the screen. Check the first Enable box and then type **xig.py** into the first Auto-start command line field, as shown in Figure 7-23.

> For XIG version 1.1.0, it's necessary to *manually* configure the Extended PAN ID for the ConnectPort's XBee radio. Click on the XBee Network link to view a list of the radios in the network (Figure 7-5). Select the gateway's radio—its node type will be listed as "coordinator"—to view its Basic Settings (Figure 7-6). Enter 0xAAAA in the Extended PAN ID field and click the Apply button.
>
> It's fine to pick any other PAN ID; just remember what it is and substitute it in the examples below. Also keep in mind that future versions of the XIG might configure the PAN ID automatically. Check the README file that comes with your download for the latest information.

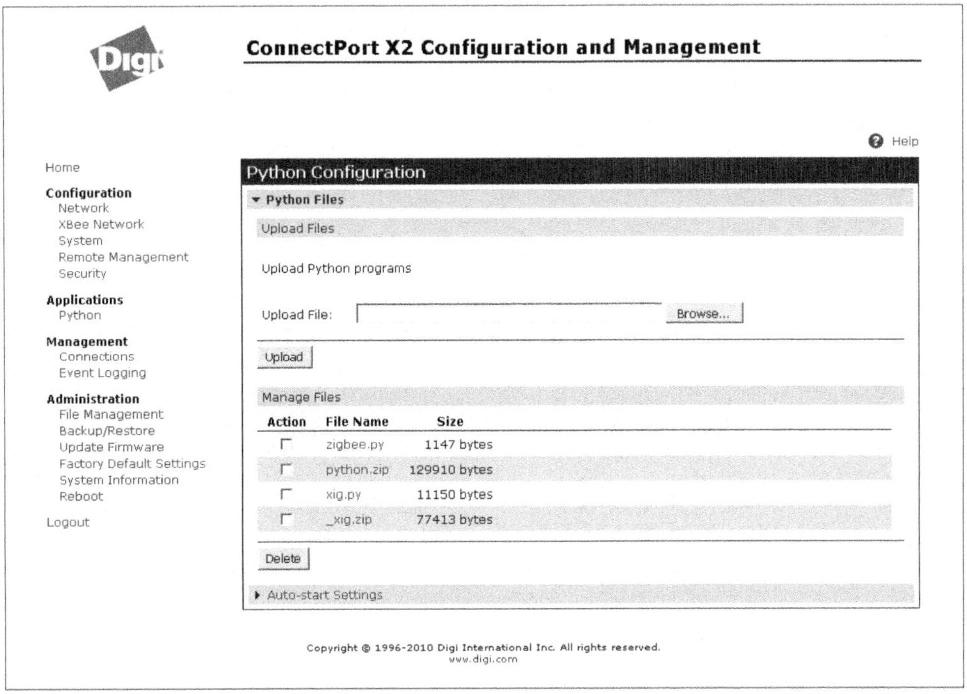

Figure 7-22. XIG files uploaded to ConnectPort

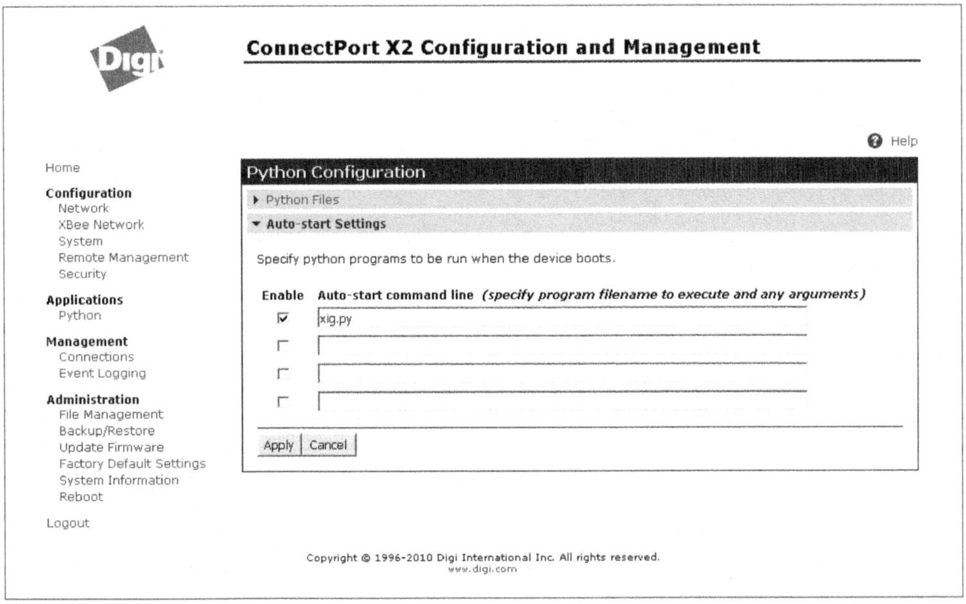

Figure 7-23. Auto-start configuration for XIG on ConnectPort

It's a good idea to set some security on your XIG. It will be connected directly to the public Internet so protecting it with a password will prevent anyone else from trying to get in and mess with your configuration. Click the Security link to enter a password (Figure 7-24). You can also change the username if you like for an extra layer of security. After you click the Apply button to make this change, you'll immediately be prompted for your user ID and password.

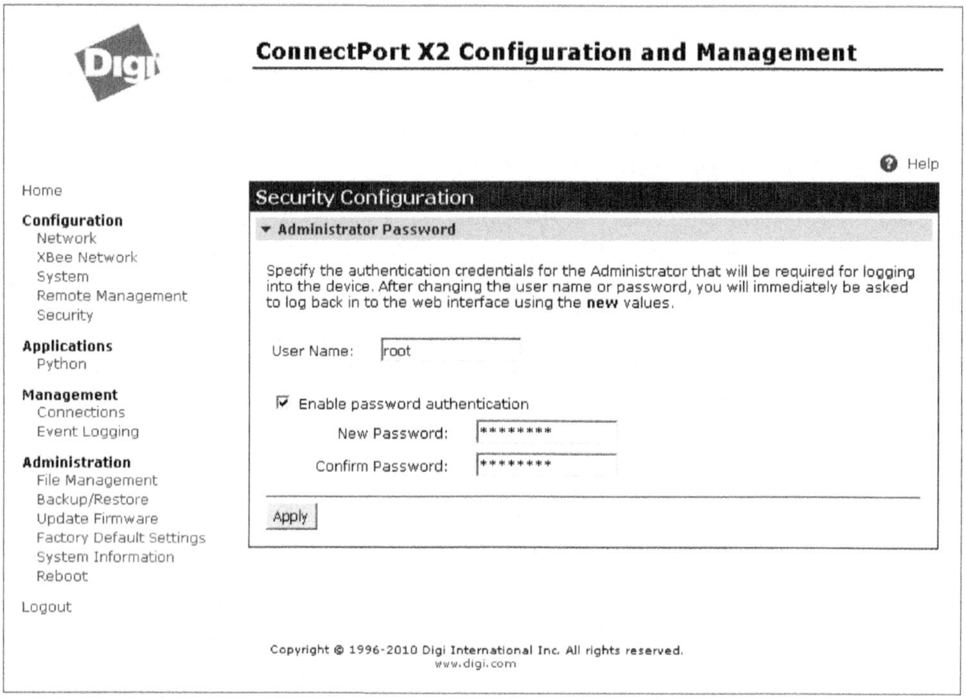

Figure 7-24. Entering security information on the ConnectPort

Finally, select the Reboot link and press the Reboot button. The XIG will now run automatically at startup.

Testing XIG

Before you start hooking up any projects, it's a good idea to confirm that your XIG is working properly. This can be done with the help of our old friend CoolTerm, or any other serial terminal program:

1. Use X-CTU to configure an XBee as a ZigBee router in AT mode.
2. Place that router into an XBee Explorer, plug it into your computer, and run the CoolTerm program.
3. Press the Options button in CoolTerm to configure the serial connection.

4. Select the appropriate serial port and check the Local Echo box.

5. Click on the Connect button to connect to the serial port.

6. Type **+++** to go into command mode.

7. Type **ATID** followed by the PAN ID. The recommended PAN for the XIG is AAAA, so **ATID AAAA** will get you configured.

8. Every ZigBee coordinator always has 0 as its 16-bit network address, and this is the default destination address for any newly configured XBee radio. Enter **ATDL 0** and **ATDH 0** to be sure that you are in the default configuration for the destination address.

9. Enter **ATJV 1** to ensure that your router attempts to rejoin the coordinator on startup.

The XIG often needs to send a lot of data, so it can be helpful to raise the baud rate. This is optional:

1. To raise the baud rate, type **ATBD** followed by the code for the baud rate you'd like to use—in this case 7 for 115,200 bps, which is the fastest speed—so type **ATBD 7** and hit Enter.

2. Once the baud rate has been raised, you'll need to select the same baud rate in CoolTerm (and later on, the same baud rate in your own project). Click Disconnect to drop the serial connection, then Options to open the settings, and switch the Baudrate setting to **115200**. Press OK and then Connect to reconnect to your XBee. It should now be responding at the higher baud rate.

To check the XIG, try typing the word **help** in the connected CoolTerm window, then pressing Return. You should get a text response with basic information on using the XIG. If that works, try entering in a URL like *http://www.faludi.com/test.html*. The HTML source for that URL should be returned, something like this:

```
<html>
<head>
<title>Rob Faludi's Test Page</title>
</head>
<body>
<p>
This is a very simple test page.
</p>
</body>
</html>
```

If you have problems, try double-checking the baud rate. You can make sure that the router XBee has joined a network by using **ATAI** and looking for a response of 0. You may also want to confirm your router is connected to the right network by issuing the **ATND** command and seeing if the gateway radio's listing comes back. Once you have confirmed everything is working, you're ready to make some stuff that's connected to the Internet!

XIG Example

PHP is a very common language for writing applications that run on web servers. A full explanation of PHP and Internet protocols is beyond the scope of this book (see *http://oreilly.com/pub/topic/php* for some books and resources on PHP). We will supply a very simple example that you can upload to your server. Use this as a starting point for creating simple URLs that your project can connect to via the XIG. On the XIG website, there is more example code that can be used to download and upload information for sensing and control.

Here's a quick look at some PHP code. Place it in a file on your server called *XIG_ download_example.php* . When you access it online, whatever value you put in the $value variable will be returned to your web browser. In this case, you'll simply see an 8.

XIG download example in PHP

```php
<?php
    // xig_download_example.php
    // this code posts a simple value in ASCII when the page is loaded
    $value = "8";
    echo $value;
?>
```

To read this in on an Arduino connected to a router XBee that you configured as described above, try the following sketch that reads in the returned value. Figure 7-25 shows a diagram of the connections. Note that it doesn't do anything with the returned value; that part is for you to write with your own purposes in mind (you must replace *<your server URL here>* with the hostname and path to wherever you uploaded *XIG_download_example.php*):

```
/*
 * *********XBee Internet Gateway Download Example********
 * by Rob Faludi http://faludi.com
 */

#define NAME "XIG Download Example"
#define VERSION "1.00"

int outputLED = 9; // define pin 9 as a PWM analog output light

void setup() {
  Serial.begin(115200); // faster is better for XIG
  pinMode(outputLED, OUTPUT);
}

void loop() {
  if (millis() % 1000 == 0) {  // wait a second before sending the next request
    // request the current value
    Serial.println("http://<your server URL here>/XIG_download_example.php");
  }
  if (Serial.available() > 0) { // if there's a byte waiting
```

```
    int value = Serial.read(); // read a single byte

    analogWrite(value, outputLED); // set an LED's brightness
                                   // to match the value
    // *** OTHER USEFUL THINGS COULD BE DONE WITH THE VALUE VARIABLE HERE ***

  }
}
```

You've seen how to download data using the XIG—a significant task that can be accomplished in a small amount of code. Examples for controlling output and uploading data are available on the XIG site that's linked from the book's website (see the Preface). Given your freshly obtained networking powers, you are probably wondering, "How can I use my new skills to display a real-time stream of celebrity gossip?" Glad you asked.

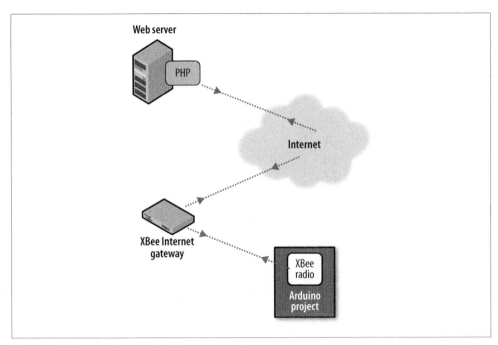

Figure 7-25. System diagram for XBee Internet Gateway connecting an Arduino project to PHP on a web server

Twitter Reader

These days, no event seems truly real until it has been reported on Twitter. From the minor details of everyday life to the tectonic shifts of the continents, an endless stream of information gushes forth from "tweeters" around the world. These missives are typically viewed on a computer screen or in a text message, but there's no reason they can't be unshackled and invited to join us off-screen out in the physical world. Why not display your Twitter feed on your office door, so people know why you aren't available? Perhaps you'd like to enjoy a feed of haikus while riding the elevator in the morning. Or maybe your tastes run a little less zen and you want to know Julia Roberts' matrimonial status as scrutinized by *US Magazine*. In all cases, the Twitter Reader example is here to help. It displays the latest message from any tweet feed wirelessly on a standard 32-character LCD display. The Reader downloads from a specially designed "twansform" online application via the XBee Internet Gateway. Twansform was written in Google App Engine by Jordan Husney. The web service used by our project can be found at *http://twansform.appspot.com* and the code is available at *http://code.google.com/p/twansform/*. It uses simple URL requests that include the feed name. In the code below the feed is *usweekly*, great for keeping tabs on the stars. Of course you can easily change it to any other account you like. For example you could use the *earthquake* feed to keep tabs on tremors, or try out *schnitznthings* if you have a hankering for flat meat in Manhattan. The Twitterverse's offerings are endless, so let's get started reading feeds.

Parts

- One solderless breadboard (large size) (AF 239, DK 438-1045-ND, RS 276-002, SFE PRT-00112)
- Hookup wire (assorted colors are particularly helpful for this project) (AF 153, DK 923351-ND, SFE PRT-00124)
- One Arduino Uno (SFE DEV-09950, AF 50) (If you use an older model, be sure it is using the new ATMEGA328 chip!)
- USB A-to-B cable for Arduino (AF 62, DK 88732-9002, SFE CAB-00512)
- Assorted 5 mm LEDs (DK 160-1707-ND, RS 276-041, SFE COM-09590)
- One 10K Ω potentiometer (panel mount) (DK P3C3103-ND, RS 271-1715, SFE COM-09288)
- One 16×2 character LCD display (with HD44780 parallel interface) (AF 181, DK 67-1758-ND, SFE LCD-00255)
- 16-pin single-row male header (generally sold in longer breakaway strips) (DK S1012E-36-ND, SFE PRT-00116)
- One ConnectPort X2 – ZB, running the XBee Internet Gateway software (Digi X2-Z11-EC-A is the new version; DK 602-1173-ND is the older version)
- One XBee radio (Series 2/ZB firmware) configured as a ZigBee Router API mode (Digi: XB24-Z7WIT-004, DK 602-1098-ND)
- One XBee breakout board with male headers and 2 mm female headers installed (AF 126 (add SFE PRT-00116), SFE BOB-08276, PRT-08272, and PRT-00116)
- XBee USB serial adapter (XBee Explorer, Digi Evaluation board, or similar) (AF 247, SFE WRL-08687)

- USB cable for XBee adapter (AF 260, SFE CAB-00598)
- Wire strippers (AF 147, DK PAL70057-ND, SFE TOL-08696)

Prepare Your ConnectPort with XBee Internet Gateway

Follow the instructions for "Installing and Configuring XIG" on page 216 to install and configure the ConnectPort X2 with the XBee Internet Gateway software. If this is a new installation, use the instructions from "Testing XIG" on page 218 to test it using a terminal program from your computer.

Prepare Your Router Radio

1. Follow the instructions under "Reading Current Firmware and Configuration" on page 35 to configure your Twitter Reader radio as a ZigBee Router AT.

 Your *router* radio will use the *AT* firmware so it can pass messages in plain text to the XIG on the ConnectPort. Be sure you select the AT version for your router!

2. Label the router radio with an "R" so that you know which one it is later on.

Configure Your Router Radio

Using the CoolTerm terminal program and an XBee Explorer USB adapter, you can set up your router radio for the Twitter Reader:

1. Select the router XBee you've labeled with an "R" and place it into the XBee Explorer.
2. Plug the XBee Explorer into your computer.
3. Run the CoolTerm program and press the Options button to configure it.
4. Select the appropriate serial port, and check the Local Echo box so you can see your commands as you type them.
5. Click on the Connect button to connect to the serial port.
6. Type **+++** to go into command mode. You should receive an OK reply from the radio.
7. Type **ATID** followed by **AAAA**, the PAN ID for the XIG on the ConnectPort, and press Enter on the keyboard. You should receive OK again as a reply.
8. Every ZigBee coordinator always has 0 as its 16-bit network address, and that's the default destination address for any newly configured XBee radio. To use 16-bit addressing, the high part of your radio's *destination* address will be zero. Type **ATDH 0** and press Enter on the keyboard. You should receive an OK response.

9. Enter **ATDL** followed by the *low* part of your radio's *destination* address, in this case also a zero because that's the fixed address for the coordinator. Type **ATDL 0** and press Enter. You should receive an OK response.

10. Enter **ATJV1** to ensure that your router attempts to rejoin a coordinator on startup.

11. Save your new settings as the radio's default by typing **ATWR** and pressing Enter.

 Tweets are very short, so for this project we are fine to avoid additional configuration steps and stick with the default serial communications rate of 9,600 baud (**ATBD3**, in case you changed it earlier). Projects that download larger datafiles will benefit from using higher baud rates.

Prepare the Twitter Reader Board

Your base station radio will use a breadboard connected to an Arduino board.

Connect power from Arduino to breadboard

1. Hook up a red wire from the 5 V output of the Arduino to one of the power rails on the breadboard. *This is a higher voltage than we used in previous projects.* The 5-volt supply is required for running the LCD screen. We will send the XBee 3.3 volts separately, directly from the Arduino as described below.

2. Hook up a black wire from either ground (GND) connection on the Arduino to a ground rail on the breadboard.

3. Hook up power and ground across the breadboard so that the rails on both sides are live.

 Make sure you are using 5 V power to supply the main breadboard.

XBee connection to Arduino

1. With the router XBee mounted on its breakout board, position the breakout board toward one end of your large breadboard so that the two rows of male header pins are inserted on opposite sides of the center trough. Leave enough free space on the breadboard for the LCD screen!

2. Use a *long* red hookup wire to connect pin 1 (VCC) of the XBee directly to the Arduino board's 3.3-volt output.

 Make sure you are supplying 3.3 V power to the XBee.

3. Use black hookup wire to connect pin 10 (GND) of the XBee to ground.

4. Use yellow (or another color) hookup wire to connect pin 2 (TX/DOUT) of the XBee to digital pin 6 on your Arduino.

 This project does not use the Arduino's hardware serial pins (0 and 1) because it employs the NewSoftSerial library. This handy library enables *any* two digital pins to be used for serial communications. Avoiding the hardware serial pins allows us to reprogram the Arduino successfully without removing any wiring. Using the NewSoftSerial library requires a little more sophisticated code, which is why we didn't do it in earlier examples. You're ready now.

5. Finally, use blue (or another color) hookup wire to connect pin 3 (RX/DIN) of your XBee to digital pin 7 on your Arduino.

Liquid crystal display (LCD) output

Tweets will be displayed on a 16-character-wide, 2-row LCD with a standard HD44780 parallel interface. These displays are very common and generally have a 16-pin interface. Displays without a backlight typically omit the last two pins. The instructions below are for backlit displays. If your LCD doesn't have one, simply ignore anything to do with pins 15 and 16. Here is a typical data sheet for a 16×2 HD44780 display: *http://www.mijnprintplaat.nl/datasheet/HD44780_16x2_Character_LCD_Display .pdf*.

1. Trim your male headers down to 16 (or 14) pins to match the number of connection holes available on your LCD. Solder the row of male headers into the LCD as shown in Figure 7-26.

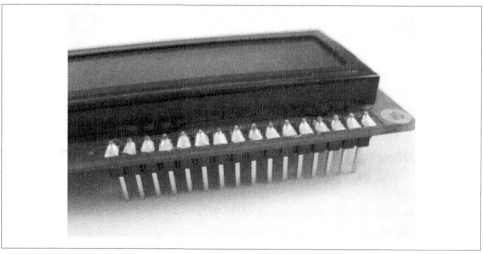

Figure 7-26. Male headers soldered to LCD

2. Insert the LCD into your breadboard. It takes up quite a bit of room, which is why you are using a larger breadboard for this project!

3. On most LCD units, the first or last pin is labeled on at least one side of its circuit board. If not, you can always consult the data sheet. Locate physical pin 1 and use a black wire to connect the LCD's physical pin 1 to one of the ground rails.

4. Use a red wire to connect the LCD's physical pin 2 to one of the power rails.

5. Attach the potentiometer to the breadboard near the LCD. Some models of potentiometer will fit right in the breadboard, while others may need jumper wires soldered onto them to make that connection happen. There are three pins, two terminals (typically the outer pins), and a wiper pin. Connect one terminal of the potentiometer to power and the other to ground. It doesn't matter which one. Connect the wiper (typically the center pin) to the LCD's physical pin 3.

 You can test your potentiometer with a multimeter to determine which pin is the wiper. Turn the potentiometer until it is about halfway between the two stops. Set your multimeter to measure resistance in ohms (Ω). A test between either terminal and the wiper will show the resistance changing as you move the knob.

6. Hook up the remaining LCD pins as shown in Table 7-1. You can also use the diagram in Figure 7-27 and the schematic in Figure 7-28 as a guide.

Table 7-1. LCD pin connections for Twitter Reader project

LCD pin #	LCD pin name	Connection
1	GND	Ground

LCD pin #	LCD pin name	Connection
2	+5V	5-volt power
3	Contrast adjustment	Potentiometer wiper
4	Register select	Arduino digital 12
5	Read/Write	Ground
6	Enable	Arduino digital 11
7	Data bus	No connection
8	Data bus	No connection
9	Data bus	No connection
10	Data bus	No connection
11	Data bus	Arduino digital 5
12	Data bus	Arduino digital 4
13	Data bus	Arduino digital 3
14	Data bus	Arduino digital 2
15	Backlight power (if available)	+5-volt power
16	Backlight GND (if available)	Ground

7. When everything is set up, plug the Arduino into your computer using the USB cable. If your LCD has a backlight, you should see it come on. You can also try adjusting the contrast on the display by turning the potentiometer so that the rectangles behind each character position just barely disappear.

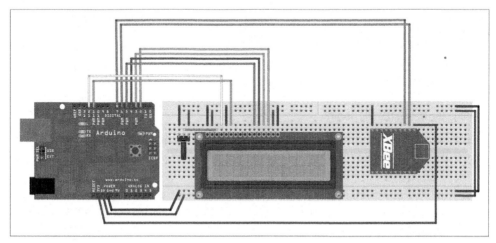

Figure 7-27. Twitter Reader breadboard layout

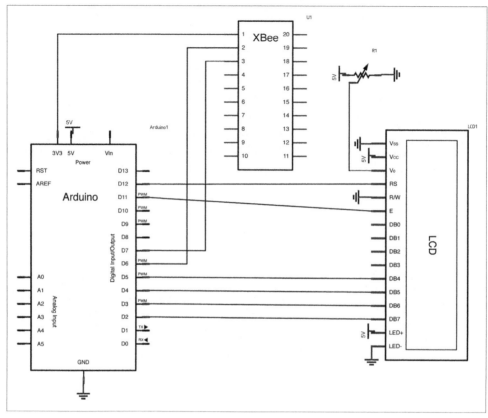

Figure 7-28. Twitter Reader schematic

Program the Arduino

The Twitter Reader uses the Arduino sketch shown later in this section. You'll also need the NewSoftSerial library.

Installing the NewSoftSerial library

Download the library from *http://arduiniana.org/libraries/newsoftserial* and unzip it. It will be in a *NewSoftSerial* folder that contains *NewSoftSerial.h*, *NewSoftSerial.cpp*, *keywords.txt*, and an *Examples* subfolder.

Open your Arduino sketchbook folder (if you're not sure where this is, open a saved sketch, choose Sketch→Show Sketch Folder, then go up to its parent directory). There is probably already a folder there called *libraries*, but if not you can create one. Place the *entire NewSoftSerial* folder inside the *libraries* folder.

If Arduino is already running, quit it and then start it up again. You should see NewSoftSerial listed on the Sketch→Import Library menu.

If you get a message like error: NewSoftSerial.h: No such file or directory when you compile your program or load it, you probably don't have a folder in the right place. Try going through the above instructions again or check *http://www.arduino.cc/en/Reference/Libraries* for more information on adding libraries to Arduino.

Once you've loaded the files and directories onto your computer, open *Twitter_Reader.pde* (full code listed below, or you can download it from the website listed in the Preface) in Arduino, press the Upload button (labeled with a right arrow) to upload the code to your Arduino. The code will run and should briefly show the words "Twitter Reader" and a version number on the LCD screen. After that the phrase loading... will be displayed while the system attempts to connect to the Internet via the XBee Internet Gateway. Here are the two lines of code that send that URL lookup request to a special Google App Engine program that parses the Twitter feed:

```
mySerial.print("http://twansform.appspot.com/usweekly/text/1");
mySerial.print("\r");
```

And here's the bit of code that reads the reply received back from the XIG into a text string:

```
// parse the incoming characters into a local String variable
char newChar;
int timeout = 4000;
while (millis()-startTime < timeout) {
  if (mySerial.available()) {
    newChar = (char)mySerial.read();
    if (newChar == '\r' || newChar == '\n') {
      break;
    }
    else {
      text.append(newChar);
    }
  }
}
```

When the lookup succeeds, you will see the latest tweet for that feed displayed. The vast majority of the program is devoted to properly displaying the tweet on the LCD, including splitting up the message properly and adding line breaks between words for maximum readability. Getting the message from the Internet is easy, thanks to the XIG! (See Figures 7-29 through 7-31.)

Figure 7-31. Twitter Reader displaying a tweet

Figure 7-29. Twitter Reader startup display

Figure 7-30. Twitter Reader shows "loading..." message while accessing the Twansform URL via XBee Internet Gateway

Twitter Reader code

```
/*
 * ********* Twitter Reader ********
 * by Rob Faludi http://faludi.com
 *
 * displays 140 characters sourced from a URL
 * using an XBee radio and a Digi ConnectPort running the XBee Internet Gateway
 * http://faludi.com/projects/xig/
 */

#include <LiquidCrystal.h>
#include <NewSoftSerial.h>
```

```
// create a software serial port for the XBee
NewSoftSerial mySerial(6, 7);
// connect to an LCD using the following pins for rs, enable, d4, d5, d6, d7
LiquidCrystal lcd(12, 11, 5, 4, 3, 2);
// defines the character width of the LCD display
#define WIDTH 16


void setup() {
  // set up the display and print the version
  lcd.begin(WIDTH, 2);
  lcd.clear();
  lcd.print("Twitter_Reader");
  lcd.setCursor(0,1);
  lcd.print("v1.04");
  delay(1000);
  lcd.clear();
  lcd.print("powered by XIG");
  lcd.setCursor(0,1);
  lcd.print("->faludi.com/xig");
  delay(2000);
  // set the data rate for the NewSoftSerial port,
  //  can be slow when only small amounts of data are being returned
  mySerial.begin(9600);
}


void loop() {
  // prepare to load some text
  String text;
  unsigned long startTime = millis();
  lcd.clear();
  lcd.print("loading...");
  // remove anything weird from the buffer
  mySerial.flush();
  // request the text string from the server
  mySerial.print("http://twansform.appspot.com/usweekly/text/1");
  mySerial.print("\r");

  // parse the incoming characters into a local String variable
  char newChar;
  int timeout = 4000;
  while (millis()-startTime < timeout) {
    if (mySerial.available()) {
      newChar = (char)mySerial.read();
      if (newChar == '\r' || newChar == '\n') {
        break;
      }
      else {
        text += newChar;
      }
    }
  }

  // clear the lcd and present the String
```

```
      if (text.length()>0) {
        unsigned long displayTime = 60000; //300000 = 5 minutes
        while(millis()-startTime < displayTime){
          lcd.clear();
          showText(text);
          // pause after showing the string
          delay(2000);
          lcd.clear();
        }
      }
    }
  }


// displays the text on an lcd with correct line breaks between words
void showText(String theText) {
  String text; // String variable for the text we are displaying
  text += theText; // puts the incoming text into our String variable
  String lineBuffer; // temporary storage for the last displayed line
  int cpos=0; // keeps track of the current cursor position
  int line=0; // keeps track of the current line
  // step through the text one character at a time
  for (int i=0; i<text.length(); i++) {
    // in general, don't make a linefeed
    boolean linefeed = false;
    if (text[i]==' ') {
      // if the current character is a space, then make a line feed
      linefeed = true;
      // ...but check first that there isn't another space before
      // the edge of the screen
      for (int j=i+1; j< i + WIDTH - cpos + 1 && j<text.length() ; j++) {
        if (text[j]==' ') linefeed=false;  // another space before
                                           // the edge of the screen
        else if (j == text.length()-1) linefeed=false; // all of the text
                                                       // completes before
                                                       // the edge of the screen
      }
    }
    // make a linefeed if we reach the edge of the screen
    // (if a word is greater in length than the width)
    if (cpos==WIDTH) {
      linefeed==true;
    }

    // on linefeeds
    if (linefeed==true) {
      switch (line) {
      case 0:
        lcd.setCursor(0,1);
        line = 1;
        break;
      case 1:
        delay(400); // brief pause at end of line
        // clear the screen
        lcd.clear();
        lcd.setCursor(0,0);
```

```
      line = 0;
      break;
    }
    cpos=0; // reset the cursor tracker to the beginning of the screen
  }

  // if this isn't a line feed
  else {
    // print the current character, add it to the line buffer and
    // advance the cursor position
    lcd.print(text[i]);
    switch (text[i]) {
    case '.':
      delay (500);
      break;
    case ',':
      delay(300);
      break;
    }
    cpos++;
    delay(100); // wait a moment after each character
  }
 }
}
```

Troubleshooting

If things don't work at first, here are some steps to try:

1. Check all your electrical connections to make sure there are no loose wires and that all the components are connected properly.

2. Check the router configuration in X-CTU to confirm that the correct modem type (XB24-ZB) and function set (ZigBee Router AT) have been selected. Also check that the PAN ID, baud rate, destination high, and destination low are configured as you expect, and that ATJV has been configured as described above.

3. An LED placed from the ASSOC pin of the reader's XBee (physical pin 15) to ground should show a flashing light.

4. Check the XBee Network screen in the ConnectPort's management window to see if it shows the Twitter Reader's XBee when you Refresh the listing. The Extended ID is the same number that is printed on the back of the XBee radio.

5. Make sure you see "Twitter Reader" and version information on your LCD when the system starts up. If not, check your connections and make sure that you have adjusted the contrast properly with the potentiometer.

6. The Twitter Reader will generally run best on an Arduino using the ATMEGA328 chipset. This will be printed on the large black microchip. You may have inconsistent results with the ATMEGA168 chipset, which only has 1K of RAM and can't always handle long strings of text.

7. We are not always able to see our own mistakes. Have a friend check everything for you. Sometimes only a second pair of eyes will catch the one or more issues that are standing in the way of success.

8. When all else fails: Try taking a break and coming back to the project after a good night's rest. Remember, many of midnight's intractable puzzles are morning's simple fixes.

Moving Forward

We're coming near the end of our fast journey through the vast world of wireless networking. You now know how to reach beyond local ZigBee environments to traverse many other systems, most notably the Internet. We've given you a great way to get started on making connections using the XBee Internet Gateway, and a full example of that connection being used in an Arduino project. The next and last chapter will give you a starting point for further expeditions into the ZigBee protocol, and outline a few advanced features that will help you in your explorations going forward. This will not be the end of the road; it's merely a stepping-off point to begin your own journey.

CHAPTER 8

More to Love

Our final chapter serves as a broader introduction to sensor networking. You are now prepared to explore advanced communication techniques, share your data, and develop your own projects. Good job! Some of the ideas and techniques we look at in this chapter are pretty technical. That's why they come here at the end, when you are in the best position to understand them. We'll start with an overview of ZigBee application messaging, examine alternate routing techniques, peruse security options, and shine a light on serial flow control. To make sure that your data doesn't languish on some local hard drive, this chapter also offers an example project for publishing your sensor network results publicly. Finally, we wrap up with a peek at the future of ZigBee and some suggestions for sharing your results with others.

Enjoyed what came before? There's more to adore. Let's go take a dip in the deep end!

Advanced ZigBee

While much of advanced ZigBee is beyond the general scope of this book, there are several things worth knowing when you need to enable larger systems or pursue inter-operation. You are now ready to meet these higher-level concepts and consider their implications for your projects. We'll discuss ZigBee stack layers, standardized APS messaging, routing scenarios, security, and serial flow control. Finally we'll point you toward reference material on the ZigBee standard.

Each of these sections is intended as a jumping-off point to get you started. Need to know more? You'll find additional detailed technical documentation in the Product Manual for the XBee ZB radios, available at: *http://www.digi.com/support/supporttype .jsp?tp=3*.

ZigBee Stack Layers

As discussed in Chapter 2, the ZigBee protocol is divided into *layers* (Figure 8-1). These are portions of the protocol that do distinct jobs, and come together to create the entire mesh communications structure in an organized and flexible fashion.

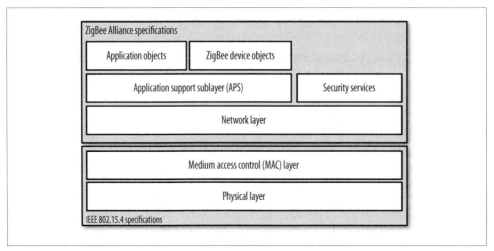

Figure 8-1. ZigBee protocol stack showing the PHY, MAC, Network, APS, and ZDO layers

We briefly covered ZigBee's physical layer (PHY) in Chapter 2, as well as the media access controller (MAC) that defines low-level addressing, among other things. Both are part of the 802.15.4 protocol that underlies ZigBee. We've also described the ZigBee Network layer that handles routing information from place to place, creating full mesh networking. There are two higher levels in the protocol that were somewhat beyond the scope of our initial forays, but may become important to you as your projects evolve and perhaps connect to other systems. They're like the frosting and decorations on the ZigBee cake:

Application Support Sublayer (APS)
> The APS layer defines standardized messaging for specific tasks where ZigBee radios are commonly employed. By creating standard messages, ZigBee devices created by different manufacturers can all carry on conversations with each other and collaborate seamlessly in predefined applications such as home automation. We will describe this layer in more detail shortly.

ZigBee Device Objects (ZDO)
> The ZDO layer is an application profile for dealing with the radios themselves. It provides device and service discovery along with certain network management capabilities. We won't discuss this layer in any more detail here and only note it for completeness. If you find yourself in a situation where requesting routing or endpoint information from a neighboring radio becomes necessary to your application,

you can learn more about it in the XBee ZB Product Manual under "ZDO Transmissions."

 These higher layers will matter when you're talking to other brands of radios, or when you're building something that needs to interact with ZigBee-certified device profiles such as Smart Energy, Home Automation, Consumer Electronics, or Health Care (see "Profiles" on page 237). If your projects only involve XBees talking to each other, then keep it simple! Read these sections purely for entertainment and don't fret about implementing the concepts.

Application Support Layer

The APS layer implements application profiles, clusters, and endpoints. You can think of these as describing the general kind of thing you're doing (*profile*), the more specific action you're taking (*cluster*), and the location within a device where that action will be carried out (*endpoint*). On the XBee, these APS messages are all sent and received by special API frames (see "Next steps" on page 239). We'll go through the concepts one by one to make it easier to understand them.

Profiles

ZigBee application profiles are collections of common definitions and protocols that allow various devices to work together in a particular domain, such as home automation. Each profile defines device types and required functionality. The most interesting profiles are the public ones that are developed and maintained by the ZigBee Alliance—the official standards organization for the ZigBee protocol—so that all ZigBee-certified devices from any manufacturer can interoperate in systems for:

- Health Care
- Home Automation
- Building Automation
- Smart Energy
- Telecommunication Services
- Consumer Electronics
- ...and new profiles that are added on a regular basis

For example, in the case of Smart Energy, the profile allows different brands of electric meters, thermostats, appliances, and in-home display units to share a common language. Any Smart Energy-certified brand of thermostat can request pricing data through any brand of certified electric meter to decide whether an extra three degrees of cooling is an affordable comfort or a pricey luxury. Similarly, any ZigBee Health Care-certified patient sensor can communicate medical data to any other brand of

Heath Care-certified patient monitor. Public profiles are available to all manufacturers who want to participate in these networks.

Manufacturers can define their own profiles internally as well. These are referred to as private profiles and are generally used within one company's products or shared in business partnerships.

 The XBee ZB communications you have been using are all part of Digi's private, manufacturer-specific Drop-in Networking profile. This is how your radios can communicate about proprietary data like analog I/O readings.

Each public or private profile has a name and a 16-bit numeric identifier to tag its messages. For example, the Smart Energy profile uses 0x0109. Any device that wants to support the Smart Energy profile is required to provide certain standard functions and to identify its messages with 0x0109 to all other devices.

 XBee-brand radios pass profile, cluster, and endpoint information inside Explicit ZigBee API frames. See "Next steps" on page 239 for references to detailed information in the Product Manual.

Every profile defines a number of clusters, which describe how two radios in a profile interact with one another. Many clusters can exist on a single endpoint. We'll look at those next.

Endpoints

In ZigBee, endpoints exist so that a device can implement multiple profiles. For example, you can have a device that belongs to both the Home Automation and Consumer Electronics profiles. It's an endpoint's job to describe applications that are running on a particular device. You can think of endpoints as mailboxes in an apartment building. To access a particular apartment, you need to know the building address but also its mailbox number. Devices frequently have many endpoints. (Readers with a basic knowledge of Internet protocols may recognize that ZigBee endpoints function similarly to port numbers in TCP/IP.) All ZigBee APS messages are sent from some endpoint on one device to some other endpoint on another device. Endpoints contain a number of clusters that are specific to a profile. Each endpoint is identified with an 8-bit number from 0x0 to 0xF0.

Clusters

In ZigBee, device profiles interact with one another through their clusters. ZigBee clusters are collections of functionality that applications can perform within a profile. Each cluster is associated with a particular action or service. For example, in the Smart Energy profile one of the clusters is devoted to price, including the various attributes and actions around setting, changing, and labeling different energy pricing scenarios and states. There are two types of clusters, client clusters and server clusters. A radio implementing a particular service—such as sending pricing information—implements the server cluster. When it has a price to transmit, it interacts with the client cluster on another radio. Likewise, client clusters may send commands that manipulate attributes or perform commands on a corresponding server cluster. Each cluster has a 16-bit ID (price messages are cluster 0x0700) and can contain various attributes and command types identified by other numeric IDs. On the XBee, for example, an API frame publishing an energy price will contain command ID 0x00 and pass a payload that includes all of the following: Provider ID, Rate Label, Issuer Event ID, Current Time, Unit of Measure, Currency, Price Trailing Digit & Price Tier, Number of Price Tiers & Register Tier, Start Time, Duration In Minutes, and finally Price. Programmers may recognize that clusters are a little like software objects, in that both contain bundles of related states and behaviors.

The ZigBee Cluster Library

Many ZigBee application profiles use a specific protocol called the ZigBee Cluster Library (or ZCL) protocol. The ZCL defines both a method for how clusters talk to one another and a collection of common actions that can be used in multiple ZigBee application profiles. For example, someone defining a new ZigBee profile for electric automobiles (a Vehicle Area Network) may import clusters of functionality from the ZCL that are already used in Home Automation or Smart Energy profiles. The ZCL exists to promote reuse, both the reuse of ideas and the reuse of source code.

The ZCL protocol encourages people to define clusters as collections of data attributes, each with a specific data type. For example, a vehicle may define a cluster of attributes for the dashboard, including the vehicle speed, the engine speed, and the amount of charge left in the battery—each defined as a whole number. The ZCL allows devices to discover which attributes are available on a cluster, read attributes, write attributes, receive periodic reports about attributes, or even receive an update when an attribute changes.

The ZCL builds on top of ZigBee clusters. Each ZCL protocol command is simply an operation sent from one cluster on one radio to another cluster on another a radio. Most commands—such as the ZCL Attribute Read command—are sent from a client cluster to a corresponding server cluster.

Next steps

Now that you know some of the basic concepts used in ZigBee APS layer communications, you can learn more by looking in the XBee ZB Product Manual (available from *http://www.digi.com/support/supporttype.jsp?tp=3*) under the API section for:

- Explicit Addressing ZigBee Command Frames

- ZigBee Explicit Rx Indicator
- Sending ZigBee Device Objects (ZDO) Commands
- Sending ZigBee Cluster Library (ZCL) Commands
- Sending Public Profile Commands with the API

Also check the latest ZigBee Specification and Public Application Profile documentation at *http://www.zigbee.org/Specifications/ZigBee/download.aspx*.

Routing

To get data from one place in the network to another, ZigBee employs several different routing methods. The first is familiar to you and is the default that's available at all times. The next two are being introduced for the first time. They must be specified using the AR command and, in some cases, special API frames:

Ad hoc On-demand Distance Vector (AODV) mesh routing
> This default method that we've been using all along automatically creates routing paths between every source and destination radio as needed. These AODV paths can hop through multiple router nodes as necessary, with every intermediate hop discovering the next step on the way to the destination address. Limitations in space for internal routing tables mean that repeated route discoveries usually need to take place to keep messages moving properly.

Many-to-one routing
> The purpose of a sensor network is often to route data messages *in* from a large number of remote nodes to one central location. Many-to-one routing is optimized for this situation. The central location broadcasts a single routing configuration message out to the network, allowing all remote devices to set up and save a reverse path back toward the central destination node. After this path is created, no more discoveries are needed for information to be properly delivered. The AR command is used to enable many-to-one broadcasting on an XBee device.

Source routing
> The purpose of other networks is to send messages from a central location *out* to multiple remote nodes. Source routing allows the central location to discover and store individual routes to a large number of remote nodes. These routes are not stored on the central location's radio, but obtained by the device or computer controlling it using the XBee's Route Record API frame. When the time comes to send a message to one particular remote node, the central location specifies a route to the remote node with the Create Source Route API frame. It then includes the data and the destination address in an API Transmit Request. Routes include the address of each intermediate hop that messages need to pass through to reach their destination.

Each routing method has its pros and cons, outlined here in Table 8-1.

Table 8-1. ZigBee routing methods compared

Routing method	Pros	Cons
AODV routing	Default method, automatically creates routes, works on any network topology.	Poor performance on large networks (more than 40 nodes) due to overhead for repeated routing requests.
Many-to-one routing	Excellent performance for multiple paths inbound to a single central location.	Not appropriate for messages outbound from a central location or networks with remote-to-remote messaging. Requires custom configuration.
Source routing	Excellent performance for outbound messages from one or more central locations, especially from highly capable devices such as computers. Good for networks of more than 40 nodes.	Requires considerable preconfiguration, including special device programming and offboard route storage. Routes must be programmatically acquired, stored, and recovered for sending.

The good news is that you won't need to worry too much about advanced routing methods on networks that are smaller than 40 nodes. However, as your networks grow, more skills with routing will become essential.

Next steps

This overview of ZigBee routing serves as a guide to learning more. You can find additional information in the XBee ZB Product Manual. Look in the Transmission, Addressing, and Routing section for RF packet routing. You'll see a detailed discussion of and instructions for:

- Link status transmission
- AODV mesh routing
- Many-to-one routing
- Source routing

In the API section look at the frame types for:

- Route record indicator
- Many-to-one route request indicator
- Create source route

Finally, check the AR, NI, and DN commands in the XBee Command Reference Tables, as these also relate to source routing. Take your time and work through your setup slowly. The concepts may not initially seem intuitive but they are very powerful and can help you build networks that are extremely large while still being remarkably efficient.

Security

ZigBee users are often anxious to employ encryption and security on their wireless networks. Security can be essential when communications or the network itself needs to be protected. Security can also be a liability, in terms of network resources and added development effort, so always consider your project. What is at risk? Do the benefits of adding security outweigh the costs? If you are transferring financial billing information, clearly your network needs all sorts of protection. On the other hand, if you are creating interactive kitty toys, adding security is only going to slow down your efforts, bog down your network, and bore your cat.

Network and link keys

ZigBee uses mathematical keys to encrypt data that is passed over the wireless network. There are two kinds of key-based security that can be used at the same time if desired: network keys and link keys.

Network keys protect your data frames as they pass between nodes. Each packet of data gets encrypted, sent to the next hop in the network, and decrypted before being reencrypted and passed along to the following node. Network security is hop-to-hop and fully protects your transmissions on private networks where all the radios are under the control of one entity. Encrypting and decrypting the packet at every hop in the route does add some transmission delay or *latency*. In addition, 18 bytes of overhead are required for the key, so data packet size is decreased from 72 to 54. This means more packets have to be sent to convey the same amount of information.

Link keys provide an added layer of end-to-end protection. Data is encrypted by the sender and remains secure as it hops along the network. Each packet is only decrypted when it reaches its destination. Use link keys to prevent intermediary hops from examining your data—useful on a shared network where individual nodes can't be trusted. For example, if you were sending personal information across a ZigBee network set up at your school, you might want to secure that data from being seen unencrypted by some radio that was legitimately a part of the network, but that you didn't control. The coordinator's link key should also be used to encrypt the distribution of the network key. Latency increases slightly and packet size is also decreased some more, so there's an additional cost to this type of layered security.

Technical details

Network security applies to data and routing messages, but not to the lower MAC-level beacon requests used by radios to first join the network. The network key is either preselected on the coordinator or set there randomly. Packets are encrypted and authenticated using 128-bit Advanced Encryption Standard (AES), a symmetric-key encryption standard adopted by the U.S. government's National Institute of Standards. A frame counter protects against replay attacks but tops out at 4 billion (32 bits). If you are going to send more than 4 billion messages (unlikely, as even at 10 messages per second it would take 13 full years), check the radio documentation for advanced suggestions about getting beyond that limit—for example, automatically leaving and rejoining the network.

APS link security (end-to-end) is also 128-bit AES. It can't be used in broadcast mode. Both network and APS link security can be used at the same time, and often are since they provide different types of protection.

Fast guide to turning on XBee network security

1. Set `ATEE` (Encryption Enable) to `1` for *all* devices on the network.
2. Set `ATNK` (Network Key) to `0` *only on the coordinator*. 0 is the default and selects a random key, which is usually fine. You could also pick a 16-position hexadecimal. The network key will be distributed to all nodes automatically. Keys are 128 bits long and the `NK` register is write-only.
3. Set `ATKY` (link KeY) to any 16-position hexadecimal (0x0000000000000001 to 0xFFFFFFFFFFFFFFFF). Use the same key for *all* radios on the network. Manually setting the coordinator's link key on each radio allows encrypted distribution of the network key. Keys are 128 bits long and the `KY` register is write-only.

Using optional APS encryption

APS end-to-end encryption can be selected on a per-packet basis by setting the enable APS options bit in the API transmit frame (see Chapter 5). Using APS encryption decreases the maximum data payload size by 9 additional bytes. This is a good example of how security is a trade-off, because it adds network latency and reduces communications efficiency. Only use it if you need it! For most, network-key-based security will be enough.

ZigBee Protocol References

If this section's taste of complexity has you hungry for more, here's where to find a cornucopia of information and documentation on the ZigBee protocol:

ZigBee Alliance Protocol Documentation
    http://www.zigbee.org/Specifications/ZigBee/download.aspx

ZigBee Alliance White Papers
    http://www.zigbee.org/LearnMore/WhitePapers.aspx

XBee Product Manuals
    http://www.digi.com/support

    http://www.digi.com/support/supporttype.jsp?tp=3

Digi White Papers
    http://www.digi.com/learningcenter/literature/whitepapers.jsp

Other ZigBee technical books
  http://www.zigbee.org/LearnMore/BooksGuides.aspx

Serial Flow Control

Throughout this book we have been using two pins on the XBee to handle serial communications. In some situations, simple TX and RX connections aren't enough to ensure your data gets delivered intact. In these cases, serial flow control can help make sure nothing is lost in transmission. Let's get some background to help understand this.

Data is held in buffers inside the XBee module during the transmit and receive process. Buffers are temporary memory locations that accumulate and hold information until it is ready to be sent to the radio antenna or serial port. They are limited in size and therefore sometimes need to be actively managed to prevent losing important data.

The serial transmit buffer inside the XBee holds data that is waiting to go out via radio over the antenna. Information is accumulated there until a full packet is ready to go or enough time has passed that the XBee decides that no more information is coming right away. Data will also wait in the transmit buffer while the XBee is receiving information, because it can't talk and listen at the same time.

The serial receive buffer holds information that is waiting to be sent from the XBee's serial port (TX pin) to the host computer or microcontroller. Computers usually can receive all the information the XBee has to send; however, in some cases microcontrollers won't always be ready to process incoming data due to limited buffer size or during times when the microcontroller program is busy doing other things.

RTS and CTS

The RTS (physical pin 16) and CTS (physical pin 12) pins on the XBee act like electronic traffic lights to control the flow of information over the RX and TX pins:

CTS
  Clear to send data *to* the XBee. When this pin is low, it is OK for the host computer or microcontroller to proceed with sending serial information. For example, you can connect this pin to one of the Arduino's digital inputs and then read its state before trying to send serial output, something like this:

```
if (digitalRead(ctsPin == LOW) ) {
  Serial.print(important_data);
}
```

  The CTS pin is de-asserted (set high) by the XBee when its serial receive buffer is almost full so that incoming data from the host doesn't overflow it and get lost.

RTS

Request to send data *from* the XBee. When the RTS pin is low, it is OK for the XBee to send serial information back to the host computer or microcontroller. The RTS pin is de-asserted (set high) by the host during times when it is not able to receive data into its own buffers. For example, you could attach one of the Arduino's digital outputs to the RTS pin, and bring the pin high anytime you are not ready to process incoming data:

```
digitalWrite(rtsPin, LOW);
if (Serial.available() > 0) {
  inVar = Serial.read();
}
//now go off do to something else
digitalWrite(rtsPin, HIGH);
```

Arduino does not currently include a native implementation of hardware flow control. These basic code examples are provided for clarity. They do not cover all the complexities you might run into when using RTS and CTS together in the same application. Check the Arduino Forums if you want to learn more: *http://arduino.cc/forum*.

Sharing Data

You've put a lot of effort into collecting data from your sensor networks. There are plenty of reasons to share the interesting information you've acquired. By putting your data someplace accessible, you'll be able to strut your stuff, share with colleagues, and possibly pipe your information into other projects all over the world. You'll also be able to monitor your own data remotely. There are a myriad of methods for sharing your data, so we'll start you off with a popular one and let that serve as your guide to self-promotion, collaboration, and cooperation.

Pachube

The Pachube site (pronounced "patch-bay") offers public upload, download, and display of data for Internet-connected sensor networks. Device data, building information, energy readings, and environmental results can be stored, shown, displayed, and downloaded from anywhere in the world. The system is free to start with as long as you make all of your data publicly available and don't upload too much at once. Privacy and extended features are available with paid accounts.

Pachube offers a wide variety of upload and download formats, including a full web services API; downloads in XML, CSV, and JSON; online graphs of various kinds; and the RSS format to track tags for new updates. Much of the data is available in Extended Environmental Markup Language, a type of XML documented at *http://eeml.org/*. You can get started with Pachube at *http://www.pachube.com/*, shown in Figure 8-2.

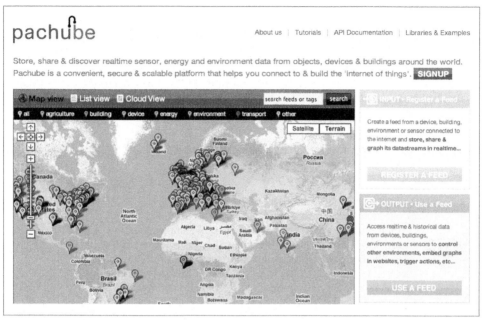

Figure 8-2. Pachube's home page, where you can sign up and register a feed

The next project modifies the simple sensor network you made in Chapter 5 to push temperature data up to Pachube for sharing. The upload code is intentionally kept very simple to show that sharing your data can be accomplished even without complex error handling, state tracking, and the like. All you need to do is create data streams and push your sensor information out to them on a regular basis.

Simple Sensor Network with Pachube

Sharing your data on the Pachube site involves setting up an account, obtaining some ID numbers, downloading an additional Processing library, and running some lightly amended Processing code that pushes your temperature network data to the site.

You'll need to sign up for a Pachube account at *http://www.pachube.com/signup* (see Figure 8-3). Accounts are free and require very little information, just a username, email address, and password to get going.

Once you are signed up, the first thing to do is register a new feed using the Register a Feed link on the home page. A feed is any collection of related data. For this project you'll use *manual* mode, meaning that your program will manually contact the Pachube site every time there is data it wants to place there. Your feed will need a title, and you can optionally enter other descriptive information as you see fit (see Figure 8-4).

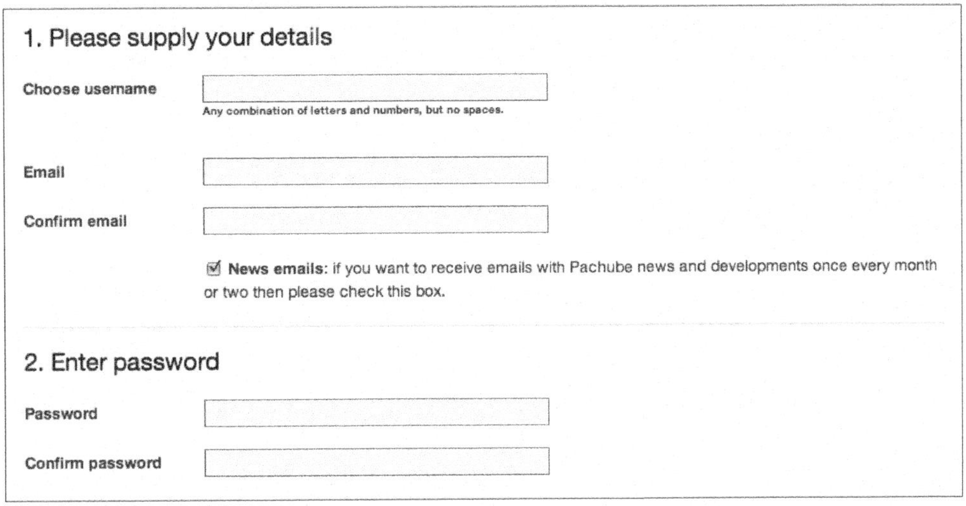

1. Please supply your details

Choose username

Any combination of letters and numbers, but no spaces.

Email

Confirm email

☑ **News emails:** if you want to receive emails with Pachube news and developments once every month or two then please check this box.

2. Enter password

Password

Confirm password

Figure 8-3. Pachube signup is free for a basic account

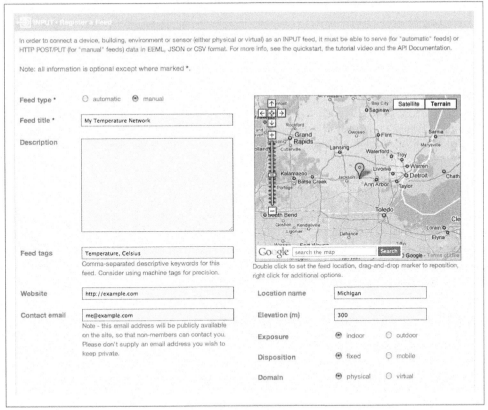

Figure 8-4. Pachube Input screen for registering a feed—any collection of related data streams

When you are done entering data, click the Save button at the bottom of the screen. This takes you to the Output window that shows you the basic feed information and several URLs for accessing data in XML, CSV, and JSON formats (Figure 8-5).

The number in each URL is your feed ID—for example *http://api.pa chube.com/v2/feeds/10298.xml*. Make a note of that feed ID number. You will need to enter it in your code for the project below!

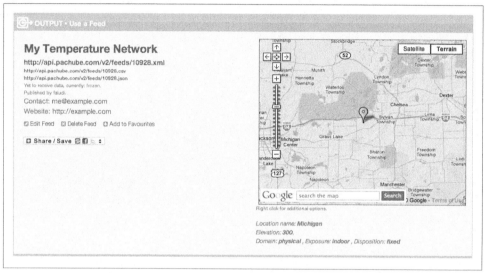

Figure 8-5. Pachube's Feed Output screen shows information about the feed, with links to receive that data in XML, CSV, or JSON formats. The number listed in the URL is the feed ID.

API Key

To access your feed remotely, Pachube requires that you pass it an API key, which is really just a long, private number to identify that it's really you uploading the data. Once you are signed in, you can get your API key by clicking on the My Profile link and selecting the Settings tab (see Figure 8-6). You'll want to copy this master API key so that you can paste it into your Processing program later.

Build the Simple Sensor Network in Chapter 5

Create the hardware and configure all the radios exactly as you did for Chapter 5's Simple Sensor Network project. Don't use the Processing code for the Chapter 5 project though! The program sketch and associated libraries in Processing are different for this Pachube version.

Figure 8-6. Pachube Master API Key screen at the My Profile→Settings tab; your personal API key must be entered in your program's code and passed on each time you update feeds manually

Program the Base Station

The simple sensor network Pachube base station uses the Processing program below. Download the ZIP file of all the libraries and resources from this book's website (see the Preface for more information). Inside the Processing sketch folder for the Simple Sensor Network are two subdirectories called *code* and *data* (see Figure 8-7). The code folder contains the *log4j.jar* and *xbee-api-0.5.5.jar* library files, just like in Chapter 5. There's also a new library here called JPachube that handles connections with Pachube. The JPachube library is available at *http://code.google.com/p/jpachube/*, where you can check for the latest version if necessary. The data folder still holds the *log4j.properties* file, required by *log4j.jar*, and the font file for the sans serif 10-point font used for screen display.

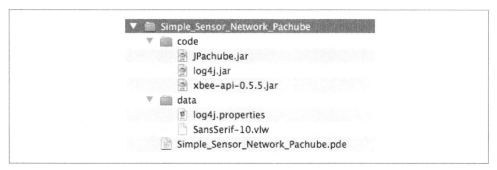

Figure 8-7. Directory structure for the Processing sketch program Simple Sensor Network Pachube, including the JPachube library used for sharing data online

You *must* replace the COM port listed in this code with your actual COM port. Look for it in the code around line 27. Port names are listed in the console in Processing, as your program starts up.

You *must also* enter your API key and the Feed ID you created in this code. Look for a string variable called apiKey and an integer variable called feedID starting near line 31.

Once you've loaded the files and directories onto your computer and opened *Simple_Sensor_Network_Pachube.pde* in Processing, press the Run button (labeled with a triangle) to launch the display code. It will open in a new window and show a thermometer for each sensor node detected. Every minute, it will attempt to upload a reading to Pachube.

If all goes well, your data will now start showing up in Pachube. Check your feed at the Pachube home page→My Feeds tab. Clicking on the feed title will show the Output screen (Figure 8-5).

The latest data will be displayed at the bottom of the Output page, with an entry for each data stream, in this case one for each temperature sensor, as shown in Figure 8-8.

ID	Tags	Value	Units	Embeddable graphs & tools
305441729	celsius temperature 00:13:A2:00:12:34:AB:C1	11.894012451171875		3 months | 4 days | **24 hours** | last hour
				reset max/min embed, history, triggers, etc
305441728	celsius temperature 00:13:A2:00:12:34:AB:C0	20.691650390625		3 months | 4 days | **24 hours** | last hour
				reset max/min embed, history, triggers, etc
305441730	celsius temperature 00:13:A2:00:12:34:AB:C2	4.152069091796875		3 months | 4 days | **24 hours** | last hour
				reset max/min embed, history, triggers, etc
Displaying all 3 datastreams				

Figure 8-8. Pachube data streams are at the bottom of the Output screen for each Feed. In this case the ID created is the decimal version of the XBee's 64-bit ID, so that each ID is unique.

Simple Sensor Network Pachube Code in Processing

Here's the source code for the Processing sketch. The comments shown in bold about the serial port, API key, and feed ID highlight essential changes. Other commented instructions are only important if you didn't download the source from the website listed in the Preface:

```
/*
 * Draws a set of thermometers for incoming XBee Sensor data
 * by Rob Faludi http://faludi.com
 */

// used for Pachube connection http://pachube.com
// JPachube library available at http://code.google.com/p/jpachube/
import Pachube.*;

// used for communication via xbee api
```

```
import processing.serial.*;

// xbee api libraries available at http://code.google.com/p/xbee-api/
// Download the zip file, extract it, and copy the xbee-api jar file
// and the log4j.jar file (located in the lib folder) inside a "code"
// folder under this Processing sketch's folder (save this sketch, then
// click the Sketch menu and choose Show Sketch Folder).
import com.rapplogic.xbee.api.ApiId;
import com.rapplogic.xbee.api.PacketListener;
import com.rapplogic.xbee.api.XBee;
import com.rapplogic.xbee.api.XBeeResponse;
import com.rapplogic.xbee.api.zigbee.ZNetRxIoSampleResponse;

String version = "1.04";

// *** REPLACE WITH THE SERIAL PORT (COM PORT) FOR YOUR LOCAL XBEE ***
String mySerialPort = "/dev/tty.usbserial-A1000iMG";

// *** REPLACE WITH YOUR OWN PACHUBE API KEY AND FEED ID ***
String apiKey="your_api_key_here";
int feedID=your_feed_id_here;

// create and initialize a new xbee object
XBee xbee = new XBee();


int error=0;



// used to record time of last data post
float lastUpdate;

// make an array list of thermometer objects for display
ArrayList thermometers = new ArrayList();
// create a font for display
PFont font;

void setup() {
  size(800, 600); // screen size
  smooth(); // anti-aliasing for graphic display


  // You'll need to generate a font before you can run this sketch.
  // Click the Tools menu and choose Create Font. Click Sans Serif,
  // choose a size of 10, and click OK.
  font = loadFont("SansSerif-10.vlw");
  textFont(font); // use the font for text

  // The log4j.properties file is required by the xbee api library, and
  // needs to be in your data folder. You can find this file in the xbee
  // api library you downloaded earlier
  PropertyConfigurator.configure(dataPath("")+"log4j.properties");
  // Print a list in case the selected one doesn't work out
  println("Available serial ports:");
```

```
    println(Serial.list());
    try {
      // opens your serial port defined above, at 9600 baud
      xbee.open(mySerialPort, 9600);
    }
    catch (XBeeException e) {
      println("** Error opening XBee port: " + e + " **");
      println("Is your XBee plugged in to your computer?");
      println(
        "Did you set your COM port in the code near line 27?");
      error=1;
    }
  }


// draw loop executes continuously
void draw() {
  background(224); // draw a light gray background
  // report any serial port problems in the main window
  if (error == 1) {
    fill(0);
    text("** Error opening XBee port: **\n"+
      "Is your XBee plugged in to your computer?\n" +
      "Did you set your COM port in the code near line 20?", width/3, height/2);
  }
  SensorData data = new SensorData(); // create a data object
  data = getData(); // put data into the data object
  //data = getSimulatedData(); // uncomment this to use random data for testing

  // check that actual data came in:
  if (data.value >=0 && data.address != null) {

    // check to see if a thermometer object already exists for this sensor
    int i;
    boolean foundIt = false;
    for (i=0; i <thermometers.size(); i++) {
      if ( ((Thermometer) thermometers.get(i)).address.equals(data.address) ) {
        foundIt = true;
        break;
      }
    }

    // process the data value into a Celsius temperature reading for
    // LM335 with a 1/3 voltage divider
    //    (value as a ratio of 1023 times max ADC voltage times
    //     3 [voltage divider value] divided by 10mV per degree
    //     minus zero Celsius in Kelvin)
    float temperatureCelsius = (data.value/1023.0*1.2*3.0*100)-273.15;
    println(" temp: " + round(temperatureCelsius) + "°C");

    // update the thermometer if it exists, otherwise create a new one
    if (foundIt) {
      ((Thermometer) thermometers.get(i)).temp = temperatureCelsius;
    }
    else if (thermometers.size() < 10) {
```

```
        thermometers.add(new Thermometer(data.address,35,450,
        (thermometers.size()) * 75 + 40, 20, data.numericAddr));
        ((Thermometer) thermometers.get(i)).temp = temperatureCelsius;
      }

    // draw the thermometers on the screen
    for (int j =0; j<thermometers.size(); j++) {
      ((Thermometer) thermometers.get(j)).render();
    }
    // post data to Pachube every minute
    if ((millis() - lastUpdate) > 60000) {
      for (int j =0; j<thermometers.size(); j++) {
        ((Thermometer) thermometers.get(j)).dataPost();
      }
      lastUpdate = millis();
    }
  }
} // end of draw loop


// defines the data object
class SensorData {
  int value;
  String address;
  long numericAddr;
}


// defines the thermometer objects
class Thermometer {
  int sizeX, sizeY, posX, posY;
  int maxTemp = 40; // max of scale in degrees Celsius
  int minTemp = -10; // min of scale in degrees Celsius
  float temp; // stores the temperature locally
  String address; // stores the address locally
  long numAddr; // stores the numeric version of the address


  Thermometer(String _address, int _sizeX, int _sizeY,
  int _posX, int _posY, long _numAddr) { // initialize thermometer object
    address = _address;
    sizeX = _sizeX;
    sizeY = _sizeY;
    posX = _posX;
    posY = _posY;
    numAddr = _numAddr;
  }

  void dataPost() {
    // add and tag a datastream
    int thermometerFeedID = (int) (numAddr);
    println("thermometerFeedID: " + thermometerFeedID);
    // initialize Pachube and Feed objects
    try {
      Pachube p = new Pachube(apiKey);
```

```
    // get the feed by its ID
    Feed f = p.getFeed(feedID);

    Data a = new Data();
    a.setId(thermometerFeedID);
    a.setMaxValue(40d);
    a.setMinValue(-10d);
    a.setTag("\"celsius temperature\", " + "\"" + address + "\"");

    // attempt to create it
    try {
      f.createDatastream(a);
    }
    catch (PachubeException e) {
      // Not a problem; this just means the feed for this
      // thermometer ID exists, and we're adding more data
      // to it now.
      if (e.errorMessage.equals("HTTP/1.1 400 Bad Request")) {
        println("feed already exists");
      }
      else {
        println(e.errorMessage);
      }
    }

    println("posting to Pachube...");
    f.updateDatastream(thermometerFeedID,
     (double) temp); // update the datastream
  }
  catch (PachubeException e) {
    println(e.errorMessage);
  }
}

void render() { // draw thermometer on screen
  noStroke(); // remove shape edges
  ellipseMode(CENTER); // center bulb
  float bulbSize = sizeX + (sizeX * 0.5); // determine bulb size
  int stemSize = 30; // stem augments fixed red bulb
  // to help separate it from moving mercury
  // limit display to range
  float displayTemp = round( temp);
  if (temp > maxTemp) {
    displayTemp = maxTemp + 1;
  }
  if ((int)temp < minTemp) {
    displayTemp = minTemp;
  }
  // size for variable red area:
  float mercury = ( 1 - ( (displayTemp-minTemp) / (maxTemp-minTemp) ));
  // draw edges of objects in black
  fill(0);
  rect(posX-3,posY-3,sizeX+5,sizeY+5);
  ellipse(posX+sizeX/2,posY+sizeY+stemSize, bulbSize+4,bulbSize+4);
  rect(posX-3, posY+sizeY, sizeX+5,stemSize+5);
```

```
    // draw gray mercury background
    fill(64);
    rect(posX,posY,sizeX,sizeY);
    // draw red areas
    fill(255,16,16);

    // draw mercury area:
    rect(posX,posY+(sizeY * mercury),
    sizeX, sizeY-(sizeY * mercury));

    // draw stem area:
    rect(posX, posY+sizeY, sizeX,stemSize);

    // draw red bulb:
    ellipse(posX+sizeX/2,posY+sizeY + stemSize, bulbSize,bulbSize);

    // show text
    textAlign(LEFT);
    fill(0);
    textSize(10);

    // show sensor address:
    text(address, posX-10, posY + sizeY + bulbSize + stemSize + 4, 65, 40);

    // show maximum temperature:
    text(maxTemp + "°C", posX+sizeX + 5, posY);

    // show minimum temperature:
    text(minTemp + "°C", posX+sizeX + 5, posY + sizeY);

    // show temperature:
    text(round(temp) + " °C", posX+2,posY+(sizeY * mercury+ 14));
  }
}

// used only if getSimulatedData is uncommented in draw loop
//
SensorData getSimulatedData() {
  SensorData data = new SensorData();
  int value = int(random(750,890));
  String address = "00:13:A2:00:12:34:AB:C" + str( round(random(0,2)) );
  data.value = value;
  data.address = address;
  data.numericAddr = unhex(data.address.replaceAll(":", ""));
  delay(200);
  return data;
}

// queries the XBee for incoming I/O data frames
// and parses them into a data object
SensorData getData() {

  SensorData data = new SensorData();
  int value = -1;      // returns an impossible value if there's an error
  String address = ""; // returns a null value if there's an error
```

```
try {
  // we wait here until a packet is received.
  XBeeResponse response = xbee.getResponse();
  // uncomment next line for additional debugging information
  //println("Received response " + response.toString());

  // check that this frame is a valid I/O sample, then parse it as such
  if (response.getApiId() == ApiId.ZNET_IO_SAMPLE_RESPONSE
      && !response.isError()) {
    ZNetRxIoSampleResponse ioSample =
      (ZNetRxIoSampleResponse)(XBeeResponse) response;

    // get the sender's 64-bit address
    int[] addressArray = ioSample.getRemoteAddress64().getAddress();
    // parse the address int array into a formatted string
    String[] hexAddress = new String[addressArray.length];
    for (int i=0; i<addressArray.length;i++) {
      // format each address byte with leading zeros:
      hexAddress[i] = String.format("%02x", addressArray[i]);
    }
    // join the array together for a numeric address:
    long numericAddress = unhex(join(hexAddress,""));
    data.numericAddr = numericAddress;
    print("numeric address: " + numericAddress);
    // join the array together with colons for readability:
    String senderAddress = join(hexAddress, ":");
    print("  sender address: " + senderAddress);
    data.address = senderAddress;
    // get the value of the first input pin
    value = ioSample.getAnalog0();
    print("  analog value: " + value );
    data.value = value;
  }
  else if (!response.isError()) {
    println("Got error in data frame");
  }
  else {
    println("Got non-i/o data frame");
  }
}
catch (XBeeException e) {
  println("Error receiving response: " + e);
}
return data; // sends the data back to the calling function
}
```

Troubleshooting

If things don't work at first, here are some steps to try:

1. Run through *all* the troubleshooting steps in Chapter 5 to ensure that the basic project's electronics and configuration are functioning properly.

2. Confirm that you entered your API key correctly in the Processing code.

3. Check that the feed ID you entered in the code matches the ID number of the feed you set up in Pachube.

4. Make sure you entered the correct port information for your XBee adapter in the Processing code.

5. Third-party services often change without warning. Check the Pachube website and the JPachube code library site to see if there have been any updates that might alter the way Pachube or the library functions.

The Future of ZigBee

One thing is for certain: ZigBee won't stay the way it is for long. The protocol is finding its way into lots of new markets so new capabilities are bound to develop quickly. One of the most interesting ventures from the ZigBee Alliance is a new protocol being developed in cooperation with the HomePlug Alliance: ZigBee Smart Energy 2.0. This new standard is broadly envisioned as a networking and application integration platform for messages between customer devices and energy services providers. The stated goal of the ZigBee+HomePlug collaboration is to "Develop a common system architecture and application profile interfaces for home energy devices, supported by a comprehensive certification process that delivers secure, robust, reliable, plug and play interoperability with AMI and Smart Grid applications."

It is reasonable to expect that innovations made in the Smart Energy 2.0 specification will inform other application profiles and, therefore, the path of ZigBee going forward. While the specifications are still being drafted, here are some of the most interesting features being discussed:

- Plans to add support for additional networking protocols, including WiFi (802.11), and HomePlug powerline. These will be in addition to the existing wireless 802.15.4. support.
- Communications over both wireless and wired networks.
- Internet Protocol addressing, including the new IPv6 standard that allows for 128-bit addresses. This creates an addressable space large enough for every device in the world to have a unique address.
- Standard UDP and TCP support so that seamless interconnections with the Internet will become possible.
- HTTP or other RESTful application protocols are expected to be included to extend these popular interconnection standards for the Web to home area devices.

This push to adopt and incorporate addressing, protocols, and interconnection standards from the global Internet should mean that direct communications between devices anywhere will become much easier to implement. Look for a future where common household objects like lamps and wall clocks can join Internet conversations as easily as a teenager can join Facebook. ZigBee stands to be a strong player in giving sensors and devices access to the world at large.

Next Steps for You

You've finally come to the end of the book, but your journey is just beginning! There's no end to the projects you can create—and plenty of excellent resources that can guide you in making your new creations.

Making Stuff

To whet your appetite, here's a list of 20 whimsically envisioned projects that could be brought into being with ZigBee radios and sensor networks. Hopefully one or more will inspire a creation of your own:

- Manage a model airplane competition.
- Make a room into a musical instrument.
- Create an electronic game of hide-and-seek.
- Monitor and display your electrical use to help reduce your bills.
- Entertain or perturb your pets.
- Keep track of open spots in a parking lot.
- Keep an eye on your grandmother's home health care.
- Link a feed of worldwide earthquakes to your massage chair.
- Network your sailboat.
- Create sock sensors to detect cold feet.
- Design reactive furniture.
- Bring a swarm of toy dinosaurs to life and simulate their migration.
- Record rainfall in an apple orchard.
- Make bracelets that sparkle when you're with your best friend.
- Create an interactive haunted house.
- Track air quality in a nearby forest.
- Send secret signals to your friends.
- Plant a garden that cares for itself.
- Sensor suit + robotic puppet = new form of ballet.
- Link a wind gauge to a fan in your cubicle and bring the outside in.
- _____ (because your own idea is the one that will change the world!).

Sharing Your Work

Networks aren't just for devices. By reading this book and creating some of its systems, you have joined a community of makers—people who include artists, engineers, crafters, scientists, hobbyists, students, teachers, entrepreneurs, hackers, and inventors. You are also one of many people using Arduino, XBee, Processing, and Python. Communities require communication and, just like networks, they work best when the whole is greater than the sum of its parts. You will find that sharing your process and projects can net feedback and recommendations from all over the world. This can make the outcome of your hard work even better. To share your projects, your code, and your hard-won wisdom:

- Participate in the forums linked from this book's website.
- On Twitter: use #BWSN to tag your tweets.
- For Instructables: use keyword BWSN.
- In Flickr: tag your photos with BWSN and add them to the BWSN group (*http:// www.flickr.com/groups/bwsn/*).
- For Pachube: tag your feed with BWSN.
- On YouTube and other video sites: tag your videos with BWSN.

Makers everywhere will be looking forward to seeing what you invented and hearing about how your projects made the journey from imagination to reality. Best of luck with your fabulous creations, and happy networking!

Resource Guide

You've savored the glamour and glitz of wireless networking; now here's some extra substance to ensure that you're fully satisfied with your high-tech meal. Since you have already launched yourself into the wireless mesh, to keep the book useful we've included links to online resources for learning more about Arduino, Processing, Python, and of course, ZigBee. There's a list of recommended books that can help you flesh out your technical library. And because every project sees its share of errors and glitches, you will find a handy troubleshooting guide to get you unstuck from common mistakes. There are tables to use as a fast daily reference of Digi radio flavors, other brands of ZigBee modules, network analyzers, packet sniffers, and XBee connectors and shields. There are also cross-referenced tables for hexadecimals, binary numbers, and ASCII codes to help keep your bytes organized, and finally, a complete guide to AT commands for the ZB radio modules.

Remember that URLs and offerings will change as time goes by, so check the book's website (listed in the Preface) for updated references to new resources.

Arduino Resources

Here are some Arduino resources you'll find useful:

Arduino Blog (http://arduino.cc/blog/)
This is the official blog, featuring Arduino news, announcements, cool projects, and more.

Arduino Forum (http://www.arduino.cc/forum/)
This is the official Arduino forum site, which contains subforums on many Arduino topics such as troubleshooting, programming, cool projects, and more.

Arduino Playground (http://www.arduino.cc/playground/)
The Playground is a wiki containing Arduino tutorials, circuits, and code. This is a good place to start if you're trying to connect Arduino to an unfamiliar device.

Ladyada.net Arduino Tutorials (http://www.ladyada.net/learn/arduino/)
    This is a great set of lessons from the folks behind Adafruit Industries (*http://www .adafruit.com/*).

Make: Online's Arduino section (http://blog.makezine.com/archive/arduino/)
    This section of Make: Online's blog is devoted to all things Arduino.

Make: Projects' Arduino section (http://makeprojects.com/Topic/Arduino)
    Make: Projects is an editable wiki full of projects.

Processing Resources

Here are some resources for the Processing language:

Processing Wiki (http://wiki.processing.org/w/Main_Page)
    This wiki features documentation, example programs, reference material, and other resources.

The Processing Feed (http://feed.processing.org/)
    Keep up-to-date on the latest news from the world of Processing.

Tutorials (http://processing.org/learning/)
    These tutorials will get you up and running with Processing and also teach you advanced Processing topics.

Processing examples (http://processing.org/learning/topics/)
    This section of the site contains many short Processing examples that can run in your browser.

Python Resources

Here are some resources for the Python language:

Python Documentation (http://www.python.org/doc/)
    This documentation covers all the currently supported versions of Python and has links to various learning resources.

Python News (http://www.python.org/news/)
    Keep up-to-date on the latest news from the Python community.

Community (http://www.python.org/community/)
    This page links to various mailing lists, wikis, user groups, and more.

The Python Tutorial (http://docs.python.org/tutorial/)
    This is the official Python tutorial.

ZigBee Resources Online

The Internet has plenty of important ZigBee resources, including specifications and information from the various standards organizations, interesting white papers, community forums, and additional tidbits compiled by other organizations.

Standards Organizations

ZigBee Alliance (http://www.zigbee.org)
> The Alliance is an association of companies that defines the protocol layers for ZigBee and promotes the standard itself as a brand.

Official ZigBee Specifications & Public Application Profiles (http://www.zigbee.org/Specifications/ZigBee/download.aspx)
> These documents contain the official definitions for how the ZigBee protocol and ZigBee public application profiles operate.

ZigBee Alliance White Papers (http://www.zigbee.org/LearnMore/WhitePapers.aspx)
> The white papers contain longer narrative explanations of application profiles, interoperation, and vision for the ZigBee project going forward.

ZigBee Alliance guide to technical books (http://www.zigbee.org/LearnMore/BooksGuides.aspx)
> This guide is a short list of technical books and resources related to ZigBee.

ZigBee Alliance guide to testing and development solutions (http://www.zigbee.org/Products/TestDevelopmentSolutions.aspx)
> Here you'll find tools and devices for testing new ZigBee products for official certification.

IEEE 802.15.4 (http://standards.ieee.org/getieee802/download/802.15.4-2006.pdf)
> This is the official specification for the network layers that live below ZigBee.

6LoWPAN ITEF (http://tools.ietf.org/html/rfc4944)
> This is the initial specification that will define ZigBee IP addressing in the future using IPv6 for low-power wireless personal area networks.

6LoWPAN document library (http://datatracker.ietf.org/wg/6lowpan)
> This library contains additional documents on 6LoWPAN.

Digi International Resources

Digi Developer Wiki (http://www.digi.com/wiki/developer)
> Here you'll find evolving documentation and examples for working with Digi products.

Digi Forums (http://www.digi.com/support/forum/listforums?category=16)
> This is the official discussion area for XBee-brand ZigBee radios.

White paper: Demystifying 802.15.4 & ZigBee (http://www.digi.com/pdf/wp_zigbee.pdf)
   Read a cogent explanation of how ZigBee is related to its underlying 802.15.4 layer.

White paper: Untangling the Mesh (http://www.digi.com/pdf/wp_untanglingthemesh .pdf)
   This guide discusses mesh networks, including wireless network basics, and gives an overview of mesh-related technologies.

White paper: Source Routing (http://www.digi.com/wiki/developer/index.php/Large_Zig Bee_Networks_and_Source_Routing)
   If you need to use the advanced source routing protocol discussed in Chapter 8, this guide can help you implement it on XBee ZB radio modules.

White paper: Antenna Considerations (http://ftp1.digi.com/support/images/XST -AN019a_XBeeAntennas.pdf)
   This is an older but still relevant guide to the different antenna options available for XBee modules.

Additional Online Resource Lists

Daintree Networks ZigBee Information (http://www.daintree.net/resources/index.php)
   This guide includes useful white papers, a comparison matrix for older versions of the ZigBee protocol, and a glossary of terms.

Palo Wireless ZigBee resources (http://www.palowireless.com/zigbee/tutorials.asp)
   This is an interesting article list that focuses on the rationale behind ZigBee and gives some comparisons to other standards, including Bluetooth.

Webcom's ZigBee Resource Guide (http://www.zigbeeresourceguide.com)
   Webcom publishes this commercially sponsored guide to ZigBee resources that includes advertising from various industry players.

ZDNet ZigBee Topics (http://www.zdnet.com/topics/zigbee)
   This is a collection of articles published by ZDNet and tagged as ZigBee-related.

Recommended Books

The following publications can help you learn more about some of the topics briefly covered in this book:

- *Programming PHP (http://oreilly.com/catalog/9780596006815/)* by Rasmus Lerdorf, et al. (O'Reilly)
- *The Visual Display of Quantitative Information* by Edward Tufte (Graphics Press)
- *The Design of Everyday Things* by Donald Norman (Basic Books)
- *Getting Started with Arduino (http://oreilly.com/catalog/9780596155520/)* by Massimo Banzi (O'Reilly)

- *Learning Processing: A Beginner's Guide to Programming Images, Animation, and Interaction* by Daniel Shiffman (Morgan Kaufmann)
- *Making Things Talk: Physical Computing with Sensors, Networks, and Arduino (http://oreilly.com/catalog/0636920010920/)* by Tom Igoe (O'Reilly)
- *Getting Started with Processing (http://oreilly.com/catalog/0636920000570/)* by Casey Reas and Ben Fry (O'Reilly)
- *Practical Electronics for Inventors* by Paul Scher (McGraw-Hill)
- *Make: Electronics (http://oreilly.com/catalog/9780596153755/)* by Charles Platt (O'Reilly)
- *Physical Computing: Sensing and Controlling the Physical World with Computers* by Tom Igoe and Dan O'Sullivan (Course Technology PTR)

Sidewalk Telescopes

At the end of the Preface, I mention sidewalk telescopes are a source of inspiration. If you'd like to build a sensor that detects the universe, here are some links to get you started:

- *http://www.telescopesineducation.com/dobson/index.html*
- *http://www.sfsidewalkastronomers.org*
- *http://quanta-gaia.org/dobson/*

Troubleshooting

When it's late at night and you just can't get your network working, more than likely a simple solution is waiting in the wings. Here are some tips for resolving issues typically encountered when working with XBee and Arduino systems.

Common XBee Mistakes

If your project won't work, check through this list of common mistakes that both beginners and experts make:

- Not using the correct firmware (choose coordinator, router, end device, and AT or API mode).
- Using ZNet 2.5 firmware, which is obsolete and will not interoperate with ZB firmware.
- Forgetting that AT commands use hexadecimals.
- Hitting Return after **+++** (or otherwise not respecting the 1-second default guard times).
- Conversely, *not* hitting Return after an AT command.

- Letting the XBee time out of command mode before issuing an AT command (you'll know because you get no response).
- Forgetting to write the configuration to firmware with ATWR (unless your application configures the radio interactively).
- Not using ATRE (restore factory defaults) before reconfiguring a previously used radio. Previous settings lurk unless you manually reset them all.
- Using a voltage regulator without decoupling capacitors (10 µF on input, 1 µF on output is usually good).
- Mixing up TX and RX pins. The fastest way to check this is to switch the wires and see if things start working.
- Trying to read more than 1.2 volts on the ZB analog inputs (1.2 V is the upper limit).
- Buying PRO radios when you don't need them. Pros cost more, are bigger, and use a lot more battery power.
- Deciding the XBees are flaky. (You may not be using them correctly; they are very reliable.)
- Deciding an XBee is burned out when it's set to a different baud rate. Check that the ON and ASSOC lights are functioning to confirm proper operation.
- Deciding an XBee is burned out when it is just sleeping. Check the ON light to see if it blinks occasionally.
- Forgetting to supply power or ground. (The ON light may go on and ASSOC light may blink, but both will be significantly dimmer.)
- Neglecting to check whether you are joined to the right network using ATAI to test for joining, and/or ATND to discover other network nodes.
- Not enabling rejoining for radios on smaller-sized networks (by setting ATJV to 1).
- Not contacting Digi sooner for support, especially if your radio seems dead or you keep getting an error you don't understand.

XBee Arduino Mistakes

Here are a few more mistakes commonly made during XBee projects that work with Arduino:

- Sending values continuously without any delay. (Try a 10 ms delay in case you are overwhelming the receiving end.)
- Not removing RX and TX connections before uploading code. (Arduino will give an error.)
- Not removing the RX connection when resetting, if you are continuously receiving data. (Arduino will never reset.)

Reference Tables

The information in this section provides a reference to radio modules, useful tools, numbering systems, and all of the XBee AT command set for the ZB radios.

Other ZigBee Modules

Although Digi's XBee radios are certainly the most popular option for certain markets, they are hardly the only option for ZigBee modules. Table A-1 shows *some* of the many manufacturers and components you might consider for your projects, including some of the ZigBee integrated circuit (IC) chips that modules use internally.

Table A-1. ZigBee module options

Manufacturer	URL	Components
Atmel	*http://www.atmel.com*	ICs, modules, development environment
California Eastern Laboratories	*http://www.cel.com/*	ICs, modules
Digi International	*http://www.digi.com*	Modules, development kits and environments
Freescale	*http://www.freescale.com*	ICs, development kits and environments
Ember	*http://www.ember.com*	ICs, development kits and environments
Jennic	*http://www.jennic.com*	ICs, modules, development environments
Laird Technologies	*http://www.lairdtech.com*	Modules, development kits
LS Research	*http://www.lsr.com/*	Modules, development kits, test environments
Microchip	*http://www.microchip.com*	Modules
Panasonic	*http://www.panasonic.com*	Modules
Radiocrafts	*http://www.radiocrafts.com*	Modules
RadioPulse	*http://www.radiopulse.co.kr*	ICs
Radiotronix	*http://www.radiotronix.com/*	Modules
Telegesis	*http://www.telegesis.com/*	Modules, development kits, USB dongles
Telit	*http://www.telit.com*	Modules, development environments, USB dongles
Texas Instruments	*http://www.ti.com*	ICs, development kits and environments

ZigBee Packet Sniffers

Table A-2 shows devices that detect and capture 802.15.4 and ZigBee radio signals for analysis. Packet sniffers are somewhat expensive but indispensable troubleshooting tools, typically used in professional network design. If you get really serious about ZigBee, you'll want to own one.

Table A-2. Network analyzers and packet sniffers

Device	Manufacturer	URL
Perytons Analyzer	Perytons	*http://www.perytons.com/products_perytonS.php*
WiSens	BzWorks Ltd.	*http://bzworks.com/wisenssoftware.htm*
ZENA Network Analyzer	Microchip Technology	*http://www.microchip.com/stellent/idcplg?IdcService=SS_GET_PAGE& nodeId=1406&dDocName=en520682*

Digi XBee Radio Modules

Table A-3 shows the various XBee radio modules that are available.

Table A-3. Guide to XBee radio module versions

XBee radio type	Protocol	Frequency	Notes
XBee ZB	ZigBee PRO	2.4 GHz	Regular and PRO high-power versions are available.
XBee ZB SMT	ZigBee PRO	2.4 GHz	Surface mount for soldering directly to printed circuit boards. Regular and PRO versions available.
XBee ZNet 2.5	ZigBee (obsolete version!)	2.4 GHz	Obsolete, but firmware can be replaced with newer ZB ZigBee PRO version in X-CTU.
XBee DigiMesh	DigiMesh	2.4 GHz	Proprietary protocol. Regular and PRO versions available.
XBee-PRO DigiMesh 900	DigiMesh	900 MHz	Proprietary protocol. Only the PRO version is available.
XBee-PRO 868	Proprietary	868 MHz	Licensed for use in Europe only.
XBee 802.15.4	IEEE 802.15.4	2.4 GHz	Regular and PRO versions are available.
XBee-PRO XSC	Proprietary	900 MHz	~10 km outdoor range.

XBee Connectors and Shields

Chapters 1 and 3 showed you some excellent options for connecting the XBee to computers via USB and directly to Arduino. Table A-4 is more comprehensive list of components that you can use to link the XBee to other devices.

Table A-4. Connectors and shields for XBee

Component	Connection	Manufacturer	URL
XBee Adapter Kit	Computer via FTDI cable, breadboard	Adafruit	*http://www.adafruit.com/index.php?main_page= product_info&products_id=126*
Arduino XBee Shield	Arduino	Arduino/Libelium	*http://arduino.cc/en/Main/ArduinoXbeeShield*
USB Development Board (part of kit)	Computer	Digi International	*http://store.digi.com/index.cfm?fuseaction=product .display&Product_ID=2352*
FIO	Arduino	Funnel	*http://funnel.cc*
XBee to USB Adapter	Computer	Gravitech	*http://store.gravitech.us/xbtousbad.html*
XBee Dongle	Computer	New Micros	*http://www.newmicros.com/cgi-bin/store/order.cgi ?form=prod_detail&part=USB-XBEE-DONGLE-CAR RIER*
USB XBee Adapter	Computer, breadboard	Parallax	*http://www.parallax.com/Store/Accessories/Commu nicationRF/tabid/161/ProductID/643/List/0/Default .aspx*
LilyPad XBee	Wearables (e.g., clothing)	SparkFun Electronics	*http://www.sparkfun.com/commerce/product_info .php?products_id=8937*
Seeeduino XBee Shield	Arduino	Seeed Studio	*http://www.seeedstudio.com/depot/xbee-shield -v11-by-seeedstudio-p-419.html*
XBee Explorer	Computer, breadboard	SparkFun Electronics	*http://www.sparkfun.com/commerce/product_info .php?products_id=8687*
XBee Explorer Serial	Computer with RS-232 serial	SparkFun Electronics	*http://www.sparkfun.com/commerce/product_info .php?products_id=9111*
SparkFun XBee Shield	Arduino	SparkFun Electronics	*http://www.sparkfun.com/commerce/product_info .php?products_id=9588*

Hex, Decimal, and Binary

Table A-5 shows the numbers 0 through 32 represented in three different numeric bases: base 16 (hexadecimal), base 10 (decimal), and base 2 (binary).

Table A-5. The numbers 0 through 32 in hex, decimal, and binary

Hexadecimal	Decimal	Binary
0x00	00	00000000
0x01	01	00000001
0x02	02	00000010
0x03	03	00000011
0x04	04	00000100
0x05	05	00000101
0x06	06	00000110
0x07	07	00000111
0x08	08	00001000
0x09	09	00001001
0x0a	10	00001010
0x0b	11	00001011
0x0c	12	00001100
0x0d	13	00001101
0x0e	14	00001110
0x0f	15	00001111
0x10	16	00010000
0x11	17	00010001
0x12	18	00010010
0x13	19	00010011
0x14	20	00010100
0x15	21	00010101
0x16	22	00010110
0x17	23	00010111
0x18	24	00011000
0x19	25	00011001
0x1a	26	00011010
0x1b	27	00011011
0x1c	28	00011100
0x1d	29	00011101
0x1e	30	00011110
0x1f	31	00011111
0x20	32	00100000

ASCII Codes

Table A-6 shows hexadecimal and decimal ASCII codes. The first 32 codes (0 through 31) are control codes used to signify changes in transmission (such as end-of-file) or special characters embedded in strings, such as tabs and line feeds. Some control codes are not in common use. These are shown in italic in the "ASCII character" column.

Table A-6. ASCII chart

Hexadecimal	Decimal	ASCII character
0x00	0	nul (null)
0x01	1	*soh* (start of heading)
0x02	2	*stx* (start of text)
0x03	3	*etx* (end of text)
0x04	4	eot (end-of-transmission or end-of-file)
0x05	5	*enq* (enquiry)
0x06	6	*ack* (acknowledge)
0x07	7	bel (bell/beep)
0x08	8	bs (backspace)
0x09	9	ht (horizontal tab)
0x0a	10	nl (line feed or newline)
0x0b	11	*vt* (vertical tab)
0x0c	12	np (form feed, page break, or new page)
0x0d	13	cr (carriage return)
0x0e	14	*so* (shift out)
0x0f	15	*si* (shift in)
0x10	16	*dle* (data link escape)
0x11	17	*dc1* (device control one)
0x12	18	*dc2* (device control two)
0x13	19	*dc3* (device control three)
0x14	20	*dc4* (device control four)
0x15	21	*nak* (negative acknowledge)
0x16	22	*syn* (synchronous idle)
0x17	23	*etb* (end transmission block)
0x18	24	*can* (cancel)
0x19	25	*em* (end-of-medium)
0x1a	26	*sub* (substitute)
0x1b	27	esc (escape)

Hexadecimal	Decimal	ASCII character
0x1c	28	*fs* (file separator)
0x1d	29	*gs* (group separator)
0x1e	30	*rs* (record separator)
0x1f	31	*us* (unit separator)
0x20	32	sp (space)
0x21	33	!
0x22	34	"
0x23	35	#
0x24	36	$
0x25	37	%
0x26	38	&
0x27	39	'
0x28	40	(
0x29	41	)
0x2a	42	*
0x2b	43	+
0x2c	44	,
0x2d	45	-
0x2e	46	.
0x2f	47	/
0x30	48	0
0x31	49	1
0x32	50	2
0x33	51	3
0x34	52	4
0x35	53	5
0x36	54	6
0x37	55	7
0x38	56	8
0x39	57	9
0x3a	58	:
0x3b	59	;
0x3c	60	<
0x3d	61	=
0x3e	62	>

Hexadecimal	Decimal	ASCII character
0x3f	63	?
0x40	64	@
0x41	65	A
0x42	66	B
0x43	67	C
0x44	68	D
0x45	69	E
0x46	70	F
0x47	71	G
0x48	72	H
0x49	73	I
0x4a	74	J
0x4b	75	K
0x4c	76	L
0x4d	77	M
0x4e	78	N
0x4f	79	O
0x50	80	P
0x51	81	Q
0x52	82	R
0x53	83	S
0x54	84	T
0x55	85	U
0x56	86	V
0x57	87	W
0x58	88	X
0x59	89	Y
0x5a	90	Z
0x5b	91	[
0x5c	92	\
0x5d	93	]
0x5e	94	^
0x5f	95	_
0x60	96	`
0x61	97	a

Hexadecimal	Decimal	ASCII character
0x62	98	b
0x63	99	c
0x64	100	d
0x65	101	e
0x66	102	f
0x67	103	g
0x68	104	h
0x69	105	i
0x6a	106	j
0x6b	107	k
0x6c	108	l
0x6d	109	m
0x6e	110	n
0x6f	111	o
0x70	112	p
0x71	113	q
0x72	114	r
0x73	115	s
0x74	116	t
0x75	117	u
0x76	118	v
0x77	119	w
0x78	120	x
0x79	121	y
0x7a	122	z
0x7b	123	{
0x7c	124		
0x7d	125	}
0x7e	126	~
0x7f	127	del (delete)

XBee Command Reference

The tables in this section describe all the commands available to you on the XBee ZB firmware as of version 2x70. They are:

Table A-7

These addressing commands allow you to specify and retrieve destinations, endpoints, parent addresses, and more. Several are discussed in Chapters 2 and 7.

Table A-8

Networking commands allow you to work with settings such as network (PAN) IDs, router configuration, and channel configuration. Chapter 2 discusses a few of these commands.

Table A-9

You can use the security settings to configure various encryption options. Security is covered in Chapter 8.

Table A-10

These RF interfacing commands let you configure power settings and retrieve the RSSI (received signal strength indication) for the last packet received.

Table A-11

Serial interfacing commands allow you to work with serial settings, switch into API mode, and more. See Chapters 7 and 8.

Table A-12

You can use the I/O commands to work with features such as PWM (will be implemented in the future versions), digital I/O, and analog input. I/O is the focus of Chapter 4.

Table A-13

These diagnostic commands let you consult the firmware and hardware versions, and also to check whether the module has associated with a network. Association indication is discussed in Chapter 7.

Table A-14

The AT command settings let you configure how the module handles AT commands that you send it. Chapter 2 discusses several of these.

Table A-15

Use these commands to configure the module's sleep mode, as covered in Chapter 6 .

Table A-16

The execution commands let you restore defaults, write the current settings to non-volatile memory, and more. Execution commands are noted throughout the book.

For each command, the "Node type" column indicates which node types support the command: C = Coordinator, R = Router, E = End Device.

Table A-7. Addressing commands

AT command	Name and description	Node type	Parameter range	Default
DH	**Destination Address High**. Sets/gets the upper 32 bits of the 64-bit destination address. When combined with DL, it defines the 64-bit destination address for data transmission. Special definitions for DH and DL include 0x000000000000FFFF (broadcast) and 0x0000000000000000 (coordinator).	CRE	0–0xFFFFFFFF	0
DL	**Destination Address Low**. Sets/gets the lower 32 bits of the 64-bit destination address. When combined with DH, it defines the 64-bit destination address for data transmissions. Special definitions for DH and DL include 0x000000000000FFFF (broadcast) and 0x0000000000000000 (coordinator).	CRE	0–0xFFFFFFFF	0xFFFF (coordinator) 0 (router/end device)
MY	**16-bit Network Address**. Reads the 16-bit network address of the module. A value of 0xFFFE means the module has not joined a ZigBee network.	CRE	0–0xFFFE (read-only)	0xFFFE
MP	**16-bit Parent Network Address**. Reads the 16-bit network address of the module's parent. A value of 0xFFFE means the module does not have a parent.	E	0–0xFFFE (read-only)	0xFFFE
NC	**Number of Remaining Children**. Reads the number of end device children that can join the device. If NC returns 0, then the device cannot allow any more end device children to join.	CR	0–MAX_CHILDREN (maximum varies)	Read-only
SH	**Serial Number High**. Reads the high 32 bits of the module's unique 64-bit address.	CRE	0–0xFFFFFFFF (read-only)	Factory-set
SL	**Serial Number Low**. Reads the low 32 bits of the module's unique 64-bit address.	CRE	0–0xFFFFFFFF (read-only)	Factory-set
NI	**Node Identifier**. Stores a string identifier. The register only accepts printable ASCII data. In AT command mode, a string cannot start with a space. A carriage return ends the command. The command will automatically end when maximum bytes for the string have been entered. This string is returned as part of the ND (Node Discover) command. This identifier is also used with the DN (Destination Node) command. In AT command mode, an ASCII comma (0x2C) cannot be used in the NI string.	CRE	20-Byte printable ASCII string	ASCII space character (0x20)

AT command	Name and description	Node type	Parameter range	Default
SE	**Source Endpoint**. Sets/reads the ZigBee application layer source endpoint value. This value will be used as the source endpoint for all data transmissions. SE is only supported in AT firmware. The default value (0xE8) is the Digi data endpoint.	CRE	0–0xFF	0xE8
DE	**Destination Endpoint**. Sets/reads ZigBee application layer destination ID value. This value will be used as the destination endpoint for all data transmissions. DE is only supported in AT firmware. The default value (0xE8) is the Digi data endpoint.	CRE	0–0xFF	0xE8
CI	**Cluster Identifier**. Sets/reads ZigBee application layer cluster ID value. This value will be used as the cluster ID for all data transmissions. CI is only supported in AT firmware. The default value (0x11) is the transparent data cluster ID.	CRE	0–0xFFFF	0x11
NP	**Maximum RF Payload Bytes**. This value returns the maximum number of RF payload bytes that can be sent in a unicast transmission. If APS encryption is used (API transmit option bit enabled), the maximum payload size is reduced by 9 bytes. If source routing is used (AR < 0xFF), the maximum payload size is reduced further. **Note**: NP returns a hexadecimal value (e.g., if NP returns 0x54, this is equivalent to 84 bytes).	CRE	0–0xFFFF	(read-only)
DD	**Device Type Identifier**. Stores a device type value. This value can be used to differentiate different XBee-based devices. Digi reserves the range 0–0xFFFFFF. For example, Digi currently uses the following DD values to identify various ZigBee products: 0x30001 - ConnectPort X8 Gateway 0x30002 - ConnectPort X4 Gateway 0x30003 - ConnectPort X2 Gateway 0x30005 - RS-232 Adapter 0x30006 - RS-485 Adapter 0x30007 - XBee Sensor Adapter 0x30008 - Wall Router 0x3000A - Digital I/O Adapter 0x3000B - Analog I/O Adapter 0x3000C - XStick	CRE	0–0xFFFFFFFF	0x30000

AT command	Name and description	Node type	Parameter range	Default
	0x3000F - Smart Plug			
	0x30011 - XBee Large Display			
	0x30012 - XBee Small Display			

Table A-8. Networking commands

AT command	Name and description	Node type	Parameter range	Default
CH	**Operating Channel**. Reads the channel number used for transmitting and receiving between RF modules. Uses 802.15.4 channel numbers. A value of 0 means the device has not joined a PAN and is not operating on any channel.	CRE	**XBee** 0, 0x0B–0x1A (Channels 11–26) **XBee-PRO (S2)** 0, 0x0B–0x18 (Channels 11–24) **XBee-PRO (S2B)** 0, 0x0B–0x19 (Channels 11–25)	(Read-only)
ID	**Extended PAN ID**. Sets/reads the 64-bit extended PAN ID. If set to 0, the coordinator will select a random extended PAN ID, and the router / end device will join any extended PAN ID. Changes to ID should be written to nonvolatile memory using the WR command to preserve the ID setting if a power cycle occurs.	CRE	0– 0xFFFFFFFFFFFFFFFF	0
OP	**Operating Extended PAN ID**. Reads the 64-bit extended PAN ID. The OP value reflects the operating extended PAN ID that the module is running on. If ID > 0, OP will equal ID.	CRE	0x01– 0xFFFFFFFFFFFFFFFF	(Read-only)
NH	**Maximum Unicast Hops**. Sets/reads the maximum hops limit. This limit sets the maximum broadcast hops value (BH) and determines the unicast timeout. The timeout is computed as (50 * NH) + 100 ms. The default unicast timeout of 1.6 seconds (NH=0x1E) is enough time for data and the acknowledgment to traverse about 8 hops.	CRE	0–0xFF	0x1E
BH	**Broadcast Hops.** Sets/reads the maximum number of hops for each broadcast data transmission. Setting this to 0 will use the maximum number of hops.	CRE	0–0x1E	0
OI	**Operating 16-bit PAN ID**. Reads the 16-bit PAN ID. The OI value reflects the actual 16-bit PAN ID the module is running on.	CRE	0–0xFFFF	(Read-only)
NT	**Node Discovery Timeout**. Sets/reads the node discovery timeout. When the network discovery (ND) command is issued,	CRE	0x20–0xFF (× 100 msec)	0x3C (60d)

AT command	Name and description	Node type	Parameter range	Default
	the NT value is included in the transmission to provide all remote devices with a response timeout. Remote devices wait a random time, less than NT, before sending their response.			
NO	**Network Discovery options**. Sets/reads the options value for the network discovery command. The options bit field value can change the behavior of the ND (network discovery) command and/or change what optional values are returned in any received ND responses or API node identification frames. Options include: 0x01 = Appends DD value (to ND responses or API node identification frames) 002 = Local device sends ND response frame when ND is issued	CRE	0–0x03 (bit field)	0

SC	**Scan Channels**. Sets/reads the list of channels to scan.
	
	Coordinator - Bit field list of channels to choose from prior to starting network.
	
	Router/End Device - Bit field list of channels that will be scanned to find a coordinator/router to join.
	
	Changes to SC should be written using the WR command to preserve the SC setting if a power cycle occurs.
	
	Bit (Channel):
	
	`0 (0x0B) 4 (0x0F)  8 (0x13)  12 (0x17)`
	`1 (0x0C) 5 (0x10)  9 (0x14)  13 (0x18)`
	`2 (0x0D) 6 (0x11)  10 (0x15)  14 (0x19)`
	`3 (0x0E) 7 (0x12)  11 (0x16)  15 (0x1A)`
		CRE	**XBee**
			1–0xFFFF (bit field)
			
			XBee-PRO (S2)
			1–0x3FFF (bit field)
			
			(bits 14, 15 not allowed)
			
			XBee-PRO (S2B)
			1–0x7FFF
			
			(bit 15 is not allowed)
				1FFE

SD	**Scan Duration**. Sets/reads the scan duration exponent. Changes to SD should be written using the WR command. **Coordinator** - Duration of the Active and Energy Scans (on each channel) that are used to determine an acceptable channel and Pan ID for the coordinator to start up on. **Router/End Device** - Duration of Active Scan (on each channel) used to locate an available coordinator/router to join during association. Scan Time is measured as: (# Channels to Scan) * (2 ∧ SD) * 15.36 ms - the number of channels to scan is determined by the SC parameter. The XBee can scan up to 16 channels (SC = 0xFFFF). Sample Scan Duration times (13 channel scan): If SD = 0, time = 0.200 sec If SD = 2, time = 0.799 sec If SD = 4, time = 3.190 sec If SD = 6, time = 12.780 sec	CRE	0–7 (exponent)	3

AT command	Name and description	Node type	Parameter range	Default
	SD influences the time the MAC listens for beacons or runs an energy scan on a given channel. The SD time is not a good estimate of the router/end device joining-time requirements. ZigBee joining adds additional overhead including beacon processing on each channel, sending a join request, etc., that extend the actual joining time.			
ZS	**ZigBee Stack Profile**. Sets/reads the ZigBee stack profile value. This must be set the same on all devices that should join the same network.	CRE	0–2	0
NJ	**Node Join Time**. Sets/reads the time that a coordinator/router allows nodes to join. This value can be changed at runtime without requiring a coordinator or router to restart. The time starts once the coordinator or router has started. The timer is reset on power-cycle or when NJ changes. For an end device to enable rejoining, NJ should be set less than 0xFF on the device that will join. If NJ < 0xFF, the device assumes the network is not allowing joining and first tries to join a network using rejoining. If multiple rejoining attempts fail, or if NJ=0xFF, the device will attempt to join using association.	CR	0–0xFF ($\times$ 1 sec)	0xFF (always allows joining)
JV	**Channel Verification**. Sets/reads the channel verification parameter. If JV=1, a router will verify the coordinator is on its operating channel when joining or coming up from a power cycle. If a coordinator is not detected, the router will leave its current channel and attempt to join a new PAN. If JV=0, the router will continue operating on its current channel even if a coordinator is not detected.	R	0 - Channel verification disabled 1 - Channel verification enabled	0
NW	**Network Watchdog Timeout**. Sets/reads the network watchdog timeout value. If NW is set > 0, the router will monitor communication from the coordinator (or data collector) and leave the network if it cannot communicate with the coordinator for three NW periods. The timer is reset each time data is received from or sent to a coordinator, or if a many-to-one broadcast is received.	R	0–0x64FF ($\times$ 1 minute) (up to over 17 days)	0 (disabled)
JN	**Join Notification**. Sets/reads the join notification setting. If enabled, the module will transmit a broadcast node identification packet on power-up and when joining. This action blinks the Association LED rapidly on all devices that receive the transmission, and sends an API frame out the UART of API devices. This feature should be disabled for large networks to prevent excessive broadcasts.	RE	0–1	0
AR	**Aggregate Routing Notification**. Sets/reads the time between consecutive aggregate route broadcast messages. If used, AR should be set on only one device to enable many-to-	CR	0–0xFF	0xFF

AT command	Name and description	Node type	Parameter range	Default
	one routing to the device. Setting AR to 0 sends only one broadcast.			

Table A-9. Security commands

AT command	Name and description	Node type	Parameter range	Default
EE	**Encryption Enable**. Sets/reads the encryption enable setting.	CRE	0 - Encryption disabled 1 - Encryption enabled	0
EO	**Encryption Options**. Configures options for encryption. Unused option bits should be set to 0. Options include: 0x01 - Send the security key unsecured over the air during joins 0x02 - Use trust center (coordinator only)	CRE	0–0xFF	--
NK	**Network Encryption Key**. Sets the 128-bit AES network encryption key. This command is write-only; NK cannot be read. If set to 0 (default), the module will select a random network key.	C	128-bit value	0
KY	**Link Key**. Sets the 128-bit AES link key. This command is write-only; KY cannot be read. Setting KY to 0 will cause the coordinator to transmit the network key in the clear to joining devices, and will cause joining devices to acquire the network key in the clear when joining.	CRE	128-bit value	0

Table A-10. RF interfacing commands

AT command	Name and description	Node type	Parameter range	Default
PL	**Power Level**. Selects/reads the power level at which the RF module transmits conducted power. For XBee-PRO (S2B) Power Level 4 is calibrated and the other power levels are approximate.	CRE	**XBee** (boost mode disabled) 0 = −8 dBm 1 = −4 dBm 2 = −2 dBm 3 = 0 dBm 4 = +2 dBm **XBee-PRO (S2)** 4 = 17 dBm **XBee-PRO (S2)** **(International Variant)**	4

AT command	Name and description	Node type	Parameter range	Default
			4 = 10 dBm	
			XBee-PRO (S2B)	
			(Boost mode enabled)	
			4 = 18 dBm	
			3 = 16 dBm	
			2 = 14 dBm	
			1 = 12 dBm	
			0 = 10 dBm	
			XBee-PRO (S2B)	
			(International Variant)	
			(Boost mode enabled)	
			4 = 10 dBm	
			3 = 8 dBm	
			2 = 6 dBm	
			1 = 4 dBm	
			0 = 2 dBm	
PM	**Power Mode**. Sets/reads the power mode of the device. Enabling boost mode will improve the receive sensitivity by 1 dB and increase the transmit power by 2 dB. **Note**: Enabling boost mode on the XBee-PRO (S2) will not affect the output power. Boost mode imposes a slight increase in current draw.	CRE	0–1, 0 = Boost mode disabled, 1 = Boost mode enabled	1
DB	**Received Signal Strength**. This command reports the received signal strength of the last received RF data packet. The DB command only indicates the signal strength of the last hop. It does not provide an accurate quality measurement for a multihop link. DB can be set to 0 to clear it. The DB command value is measured in −dBm. For example, if DB returns 0x50, then the RSSI of the last packet received was −80dBm. As of 2x6x firmware, the DB command value is also updated when an APS acknowledgment is received.	CRE	0–0xFF Observed range for XBee-PRO: 0x1A–0x58 For XBee: 0x1A–0x5C	
PP	**Peak Power**. Reads the dBm output when maximum power is selected (PL4).	CRE	0x0–0x12	(Read-only)

Table A-11. Serial interfacing commands

AT command	Name and description	Node type	Parameter range	Default
AP	**API Enable**. Enables API mode. The AP command is only supported when using API firmware: 21xx (API coordinator), 23xx (API router), and 29xx (API end device).	CRE	1–2 1 = API-enabled 2 = API-enabled (w/ escaped control characters)	1
AO	**API Options**. Configures options for API. Current options select the type of receive API frame to send out the UART for received RF data packets.	CRE	0 - Default receive API indicators enabled 1 - Explicit Rx data indicator API frame enabled (0x91) 3 - Enable ZDO pass-through of ZDO requests to the UART, which are not supported by the stack, as well as Sim ple_Desc_req, Active_EP_req, and Match_Desc_req.	0
BD	**Interface Data Rate**. Sets/reads the serial interface data rate for communication between the module serial port and host. Any value above 0x07 will be interpreted as an actual baud rate. When a value above 0x07 is sent, the closest interface data rate represented by the number is stored in the BD register.	CRE	0–7 (standard baud rates) 0 = 1,200 bps 1 = 2,400 2 = 4,800 3 = 9,600 4 = 19,200 5 = 38,400 6 = 57,600 7 = 115,200 0x80–0xE1000 (nonstandard rates up to 921 kbps)	3
NB	**Serial Parity**. Sets/reads the serial parity setting on the module.	CRE	0 = No parity 1 = Even parity 2 = Odd parity 3 = Mark parity	0
SB	**Stop Bits**. Sets/reads the number of stop bits for the UART. (Two stop bits are not supported if mark parity is enabled.)	CRE	0 = 1 stop bit 1 = 2 stop bits	0
RO	**Packetization Timeout**. Sets/reads number of character times of intercharacter silence required before packetization. Set	CRE	0–0xFF (× character times)	3

AT command	Name and description	Node type	Parameter range	Default
	(RO=0) to transmit characters as they arrive instead of buffering them into one RF packet. The RO command is only supported when using AT firmware: 20xx (AT coordinator), 22xx (AT router), and 28xx (AT end device).			
D7	**DIO7 Configuration**. Selects/reads options for the DIO7 line of the RF module.	CRE	0 = Disabled	1
			1 = CTS flow control	
			3 = Digital input	
			4 = Digital output, low	
			5 = Digital output, high	
			6 = RS-485 transmit enable (low enable)	
			7 = RS-485 transmit enable (high enable)	
D6	**DIO6 Configuration**. Configures options for the DIO6 line of the RF module.	CRE	0 = Disabled	0
			1 = RTS flow control	
			3 = Digital input	
			4 = Digital output, low	
			5 = Digital output, high	

Table A-12. I/O commands

AT command	Name and description	Node type	Parameter range	Default
IR	**IO Sample Rate**. Sets/reads the IO sample rate to enable periodic sampling. For periodic sampling to be enabled, IR must be set to a nonzero value, and at least one module pin must have analog or digital IO functionality enabled (see D0–D8, P0–P2 commands). The sample rate is measured in milliseconds.	CRE	0, 0x32:0xFFFF (ms)	0
IC	**IO Digital Change Detection**. Sets/reads the digital IO pins to monitor for changes in the IO state. IC works with the individual pin configuration commands (D0–D8, P0–P2). If a pin is enabled as a digital input/output, the IC command can be used to force an immediate IO sample transmission when the DIO state changes. IC is a bit mask that can be used to enable or disable edge detection on individual channels. Unused bits should be set to 0. Bit (IO pin):	CRE	0–0xFFFF	0

AT command	Name and description	Node type	Parameter range	Default
	0 (DIO0) 4 (DIO4) 8 (DIO8)			
	1 (DIO1) 5 (DIO5) 9 (DIO9)			
	2 (DIO2) 6 (DIO6) 10 (DIO10)			
	3 (DIO3) 7 (DIO7) 11 (DIO11)			
P0	**PWM0 Configuration**. Selects/reads function for PWM0.	CRE	0 - Disabled	1
			1 - RSSI PWM	
			3 - Digital input, monitored	
			4 - Digital output, default low	
			5 - Digital output, default high	
P1	**DIO11 Configuration**. Configures options for the DIO11 line of the RF module.	CRE	0 - Unmonitored digital input	0
			3 - Digital input, monitored	
			4 - Digital output, default low	
			5 - Digital output, default high	
P2	**DIO12 Configuration**. Configures options for the DIO12 line of the RF module.	CRE	0 - Unmonitored digital input	0
			3 - Digital input, monitored	
			4 - Digital output, default low	
			5 - Digital output, default high	
P3	**DIO13 Configuration**. Sets/reads function for DIO13. This command is not yet supported.	CRE	0, 3–5	--
			0 – Disabled	
			3 – Digital input	
			4 – Digital output, low	
			5 – Digital output, high	
D0	**AD0/DIO0 Configuration**. Selects/reads function for AD0/DIO0.	CRE	1 - Commissioning button enabled	1
			2 - Analog input, single-ended	
			3 - Digital input	
			4 - Digital output, low	
			5 - Digital output, high	
D1	**AD1/DIO1 Configuration**. Selects/reads function for AD1/DIO1.	CRE	0, 2–5	0
			0 – Disabled	
			2 - Analog input, single-ended	
			3 - Digital input	
			4 - Digital output, low	

AT command	Name and description	Node type	Parameter range	Default
			5 - Digital output, high	
D2	**AD2/DIO2 Configuration**. Selects/reads function for AD2/DIO2.	CRE	0, 2–5	0
			0 - Disabled	
			2 - Analog input, single-ended	
			3 - Digital input	
			4 - Digital output, low	
			5 - Digital output, high	
D3	**AD3/DIO3 Configuration**. Selects/reads function for AD3/DIO3.	CRE	0, 2-5	0
			0 - Disabled	
			2 - Analog input, single-ended	
			3 - Digital input	
			4 - Digital output, low	
			5 - Digital output, high	
D4	**DIO4 Configuration**. Selects/reads function for DIO4.	CRE	0, 3–5	0
			0 - Disabled	
			3 - Digital input	
			4 - Digital output, low	
			5 - Digital output, high	
D5	**DIO5 Configuration**. Configures options for the DIO5 line of the RF module.	CRE	0 - Disabled	1
			1 - Associated	
			indication LED	
			3 - Digital input	
			4 - Digital output, default low	
			5 - Digital output, default high	
D8	**DIO8 Configuration**. Sets/reads function for DIO8. This command is not yet supported.	CRE	0, 3–5	
			0 - Disabled	
			3 - Digital input	
			4 - Digital output, low	
			5 - Digital output, high	
LT	**Assoc LED Blink Time**. Sets/reads the Association LED blink time. If the Association LED functionality is enabled (D5 command), this value determines the on and off blink times for the LED when the module has joined a network. If LT=0, the default	CRE	0, 0x0A–0xFF (100–2,550 ms)	0

AT command	Name and description	Node type	Parameter range	Default
	blink rate will be used (500 ms for coordinator, 250 ms for router/end device). For all other LT values, LT is measured in 10 ms.			
PR	**Pull-up Resistor**. Sets/reads the bit field that configures the internal pull-up resistor status for the I/O lines. "1" specifies the pull-up resistor is enabled. "0" specifies no pull-up. (30k pull-up resistors.) Bits: 0 - DIO4 (Pin 11) 1 - AD3 / DIO3 (Pin 17) 2 - AD2 / DIO2 (Pin 18) 3 - AD1 / DIO1 (Pin 19) 4 - AD0 / DIO0 (Pin 20) 5 - RTS / DIO6 (Pin 16) 6 - DTR / Sleep Request / DIO8 (Pin 9) 7 - DIN / Config (Pin 3) 8 - Associate / DIO5 (Pin 15) 9 - On/Sleep / DIO9 (Pin 13) 10 - DIO12 (Pin 4) 11 - PWM0 / RSSI / DIO10 (Pin 6) 12 - PWM1 / DIO11 (Pin 7) 13 - CTS / DIO7 (Pin 12)	CRE	0–0x3FFF	0–0x1FFF
RP	**RSSI PWM Timer**. Number of times the RSSI signal will be output on the PWM after the last RF data reception or APS acknowledgment. When RP = 0xFF, output will always be on.	CRE	0–0xFF (× 100 ms)	0x28 (40d)
%V	**Supply Voltage**. Reads the voltage on the Vcc pin. Scale by 1200/1024 to convert to mV units. For example, a %V reading of 0x900 (2,304 decimal) represents 2,700 mV or 2.70V.	CRE	−0x–0xFFFF (read-only)	--
V+	**Voltage Supply Monitoring**. The voltage supply threshold is set with the V+ command. If the measured supply voltage falls below or equal to this threshold, the supply voltage will be included in the IO sample set. V+ is set to 0 by default (do not include the supply voltage). Scale mV units by 1,024/1,200 to convert to internal units. For example, for a 2,700 mV threshold enter 0x900.	CRE	0–0xFFFF	0

AT command	Name and description	Node type	Parameter range	Default
	Given the operating Vcc ranges for different platforms, and scaling by 1,024/1,200, the useful parameter ranges are:			
	XBee 2,100–3,600 mV, 0,0x0700–0x0c00			
	PRO 3,000–3,400 mV, 0,0x0a00–0x0b55			
	S2B 2,700–3,600 mV, 0,0x0900–0x0c00			
TP	Reads the module temperature in degrees Celsius. Accuracy +/− 7 degrees.	CRE	0x0–0xFFFF	--
	1°C = 0x0001 and −1°C = 0xFFFF. Command is only available in PRO S2B.			

Table A-13. Diagnostics commands

AT command	Name and description	Node type	Parameter range	Default
VR	**Firmware Version**. Reads firmware version of the module.	CRE	0–0xFFFF (read-only)	Factory-set
	The firmware version returns 4 hexadecimal values (2 bytes) "ABCD". Digits "ABC" are the main release number and "D" is the revision number from the main release. "B" is a variant designator.			
	XBee and XBee-PRO ZB modules return:			
	0x2xxx versions.			
	XBee and XBee-PRO ZNet modules return:			
	0x1xxx versions. ZNet firmware is not compatible with ZB firmware.			
HV	**Hardware Version**. Reads the hardware version of the module. This command can be used to distinguish among different hardware platforms. The upper byte returns a value that is unique to each module type. The lower byte indicates the hardware revision.	CRE	0–0xFFFF (read-only)	Factory-set
	XBee ZB and XBee ZNet modules return the following (hexadecimal) values:			
	0x19xx - XBee module			
	0x1Axx - XBee-PRO module			
AI	**Association Indication**. Reads information regarding last node join request:	CRE	0–0xFF (read-only)	--
	0x00 - Successfully formed or joined a network. (Coordinators form a network, routers and end devices join a network.)			
	0x21 - Scan found no PANs.			
	0x22 - Scan found no valid PANs based on current SC and ID settings.			

AT command	Name and description	Node type	Parameter range	Default
	0x23 - Valid coordinator or routers found, but they are not allowing joining (NJ expired).			
	0x24 - No joinable beacons were found.			
	0x25 - Unexpected state; node should not be attempting to join at this time.			
	0x27 - Node joining attempt failed (typically due to incompatible security settings).			
	0x2A - Coordinator start attempt failed.			
	0x2B - Checking for an existing coordinator.			
	0x2C - Attempt to leave the network failed.			
	0xAB - Attempted to join a device that did not respond.			
	0xAC - Secure join error—network security key received unsecured.			
	0xAD - Secure join error—network security key not received.			
	0xAF - Secure join error—joining device does not have the right preconfigured link key.			
	0xFF - Scanning for a ZigBee network (routers and end devices).			
	Note: New nonzero AI values may be added in later firmware versions. Applications should read AI until it returns 0x00, indicating a successful startup (coordinator) or join (routers and end devices).			

Table A-14. AT Command Options commands

AT command	Name and description	Node type	Parameter range	Default
CT	**Command Mode Timeout**. Sets/reads the period of inactivity (no valid commands received) after which the RF module automatically exits AT command mode and returns to idle mode.	CRE	2–0x028F (× 100 ms)	0x64 (100d)
CN	**Exit Command Mode**. Explicitly exits the module from AT command mode.	CRE	--	--
GT	**Guard Times**. Sets required period of silence before and after the Command Sequence Characters of the AT Command Mode Sequence (GT + CC + GT). The period of silence is used to prevent inadvertent entrance into AT command mode.	CRE	1–0x0CE4 (× 1 ms) (max of 3.3 decimal sec)	0x3E8 (1,000 d)
CC	**Command Sequence Character**. Sets/reads the ASCII character value to be used between Guard Times of the AT Command Mode Sequence (GT + CC + GT). The AT Command Mode Sequence enters the RF module into AT command mode. The CC command is only supported when using AT firmware: 20xx (AT coordinator), 22xx (AT router), and 28xx (AT end device).	CRE	0 –0xFF	0x2B ('+' ASCII)

Table A-15. Sleep commands

AT command	Name and description	Node type	Parameter range	Default
SM	**Sleep Mode**. Sets the sleep mode on the RF module. An XBee loaded with router firmware can be configured as either a router (SM set to 0) or an end device (SM > 0). Changing a device from a router to an end device (or vice versa) forces the device to leave the network and attempt to join as the new device type when changes are applied.	RE	0 - Sleep disabled (router) 1 - Pin sleep enabled 4 - Cyclic sleep enabled 5 - Cyclic sleep, pin wake	0 - Router 4 - End device
SN	**Number of Sleep Periods**. Sets the number of sleep periods to not assert the On/Sleep pin on wake-up if no RF data is waiting for the end device. This command allows a host application to sleep for an extended time if no RF data is present.	CRE	1–0xFFFF	1
SP	**Sleep Period**. This value determines how long the end device will sleep at a time, up to 28 seconds. (The sleep time can effectively be extended past 28 seconds using the SN command.) On the parent, this value determines how long the parent will buffer a message for the sleeping end device. It should be set at least equal to the longest SP time of any child end device.	CRE	0x20–0xAF0 (× 10 ms) (Quarter-second resolution)	0x20
ST	**Time Before Sleep**. Sets the time-before-sleep timer on an end device. The timer is reset each time serial or RF data is received. Once the timer expires, an end device may enter low-power operation. Applicable for cyclic sleep end devices only.	E	1–0xFFFE (× 1 ms)	0x1388 (5 seconds)
SO	**Sleep Options**. Configures options for sleep. Unused option bits should be set to 0. Sleep options include: 0x02 - Always wake for ST time 0x04 - Sleep entire SN * SP time Sleep options should not be used for most applications. See Chapter 6 for more information.	E	0–0xFF	0
WH	**Wake Host**. Sets/reads the wake host timer value. If the wake host timer is set to a nonzero value, this timer specifies a time (in millisecond units) that the device should allow after waking from sleep before sending data out the UART or transmitting an IO sample. If serial characters are received, the WH timer is stopped immediately.	E	0–0xFFFF (× 1 ms)	
SI	**Sleep Immediately**. See Table A-16.			
PO	**Polling Rate**. Sets/reads the end device poll rate. Setting this to 0 (default) enables polling at 100 ms (default rate). Adaptive polling may allow the end device to poll more rapidly for a short time when receiving RF data.	E	0–0x3E8	0x00 (100 msec)

AT command	Name and description	Node type	Parameter range	Default
AC	**Apply Changes**. Applies changes to all command registers, causing queued command register values to be applied. For example, changing the serial interface rate with the BD command will not change the UART interface rate until changes are applied with the AC command. The CN command and 0x08 API command frame also apply changes.	CRE	--	--
WR	**Write**. Writes parameter values to nonvolatile memory so that parameter modifications persist through subsequent resets.	CRE	--	--
	Note: Once WR is issued, no additional characters should be sent to the module until after the "OK\r" response is received. The WR command should be used sparingly. The EM250 supports a limited number of write cycles.			
RE	**Restore Defaults**. Restores module parameters to factory defaults.	CRE	--	--
FR	**Software Reset**. Resets module. Responds immediately with an OK status, and then performs a software reset about 2 seconds later.	CRE	--	--
NR	**Network Reset**. Resets network layer parameters on one or more modules within a PAN. Responds immediately with an "OK," then causes a network restart. All network configuration and routing information is consequently lost.	CRE	0–1	--
	If NR = 0: Resets network layer parameters on the node issuing the command.			
	If NR = 1: Sends broadcast transmission to reset network layer parameters on all nodes in the PAN.			
SI	**Sleep Immediately**. Causes a cyclic sleep module to sleep immediately rather than wait for the ST timer to expire.	E	--	--
CB	**Commissioning Pushbutton**. This command can be used to simulate commissioning button presses in software. The parameter value should be set to the number of button presses to be simulated. For example, sending the ATCB1 command will execute the action associated with one commissioning button press.	CRE	--	--

ND	**Node Discover**. Discovers and reports all RF modules found. The following information is reported for each module discovered:
	
	`MY<CR>`
	  `SH<CR>`
	  `SL<CR>`
	  `NI<CR> (Variable length)`
	  `PARENT_NETWORK ADDRESS (2 Bytes)<CR>`
	  `DEVICE_TYPE<CR> (1 Byte:`
	    `0=Coord, 1=Router, 2=End Device)`
	  `STATUS<CR> (1 Byte: Reserved)`
	  `PROFILE_ID<CR> (2 Bytes)`
	  `MANUFACTURER_ID<CR> (2 Bytes)`
	  `<CR>`
		CRE	Optional 20-byte NI or MY value	--

AT command	Name and description	Node type	Parameter range	Default
	After (NT * 100) milliseconds, the command ends by returning a <CR>. ND also accepts a Node Identifier (NI) as a parameter (optional). In this case, only a module that matches the supplied identifier will respond.			
	If ND is sent through the API, each response is returned as a separate AT_CMD_Response packet. The data consists of the above-listed bytes without the carriage return delimiters. The NI string will end in a "0x00" null character. The radius of the ND command is set by the BH command.			
DN	**Destination Node**. Resolves an NI (Node Identifier) string to a physical address (case-sensitive). The following events occur after the destination node is discovered: AT Firmware 1. DL and DH are set to the extended (64-bit) address of the module with the matching NI (Node Identifier) string. 2. OK (or ERROR)\r is returned. 3. Command mode is exited to allow immediate communication. API Firmware The 16-bit network and 64-bit extended addresses are returned in an API Command Response frame. If there is no response from a module within (NT * 100) milliseconds or a parameter is not specified (left blank), the command is terminated and an "ERROR" message is returned. In the case of an ERROR, command mode is not exited. The radius of the DN command is set by the BH command.	CRE	Up to 20-byte printable ASCII string	--
IS	**Force Sample**. Forces a read of all enabled digital and analog input lines.	CRE	--	--
1S	**XBee Sensor Sample**. Forces a sample to be taken on an XBee sensor device. This command can only be issued to an XBee sensor device using an API remote command.	RE	--	--

Index

Symbols

16-bit addressing, 29, 126
3G and 4G protocol, 191
64-bit addressing, 28, 126
128-bit Advanced Encryption Standard (AES), 242

A

accelerometer, 86
acoustic sensor, 86, 87
actuation, xii
    (see also direct actuation; remote actuation)
ad hoc network creation, 26
Ad hoc On-demand Distance Vector (AODV) mesh routing, 240
AD0...AD3 Analog Input pins, 15, 89
Adafruit electronics kits, 63
Adafruit XBee Adapter Kit, 8
adapter
    Arduino board for, 12–15
    breakout board, 10–11
    buying, 6–15
    Digi evaluation board, 7
    drivers for, 8, 33
    ports for, 40
    USB adapter, 7–9
addresses, network, 28–29
addressing commands, 276–278
AES (Advanced Encryption Standard), 242
Analog Input pins (AD0...AD3), 15, 89
animism, x
antennas, 4–5
AODV (Ad hoc On-demand Distance Vector) mesh routing, 240

API (application programming interface), 111–112, 116–119
API frame, 117–119
    AT Command frame, 120–122
    AT Response frame, 122–124
    checksum, 118
    data bytes, 118
    I/O Data Sample Rx Indicator frame, 131–135
    length bytes, 118
    parsing, API code for, 138–140
    parsing, libraries for, 141
    Remote AT Command Request frame, 135–137
    Remote Command Response frame, 137–138
    start delimiter, 117
    types of, 119–142
    ZigBee Receive Packet frame, 129–131
    ZigBee Transmit Request frame, 124–127
    ZigBee Transmit Status frame, 127–129
API protocol, 116–119
application layer, 30
application profiles, 237–238
application programming interface (see API)
APS (Application Support Sublayer) layer, 236, 237–240
APS encryption, 243
APS link security, 243
AR command, 240
Arduino & C/C++ library, 141
Arduino board, 57–65
    adapter hack for, 12–15
    buying, 59
    cable for, 59, 60

We'd like to hear your suggestions for improving our indexes. Send email to *index@oreilly.com*.

ATSO (Sleep Options), 166, 167, 290
ATSP (Sleep Period), 164, 166, 290
ATST (Time Before Sleep), 165, 166, 290
ATTP (Module Temperature), 288
ATV+ (Voltage Supply Monitoring), 287
ATVR (Firmware Version), 288
ATWH (Wake Host), 166, 167, 290
ATWR (Write), 48, 91, 291
ATZS (ZigBee Stack Profile), 280

B

battery life, determining, 163
Bluetooth protocol, 189, 190
board-level serial (see TTL serial protocol)
books, recommended, 264
breadboard, 9
breakout board, 10–11
bytes, 114, 135

C

California Eastern Laboratories, 267
CAN (Controller-Area Networking) protocol, 190
CAN-bus protocol, 190
capacitance sensor, 86
cellular data connections, 194
channels, network, 29
chat session, 50–56
    addresses for, 50
    coordinator for, 51
    with one computer, 54
    parts for, 50
    router for, 53
    with two computers, 53
checksum, 116, 118
cluster tree network, 28
clusters, 238–239
code examples, permission to use, xiii
color sensor, 86
command mode, 43, 44, 46–47
COMMISioning pin, 15
communication settings, terminal program, 40
computer, as Internet gateway, 195
ConnectPort gateways, 195–202
    configuring, 198–202
    configuring on iDigi server, 209–210
    connecting to iDigi server, 206–209

firmware updates using iDigi server, 210–212
    remote management of, 203–214
    setting up, 197–198
contact information for this book, xiv
contents ID, in stream, 115–116
context phenomena, 87
Controller-Area Networking (CAN) protocol, 190
conventions used in this book, xiii
CoolTerm program, 18, 43–46
coordinator device, 26
CTS pin, 15, 244

D

data
    presentation on Internet, 193
    reasons for collecting, 85
    sharing, 245–257
    storage on Internet, 193
data bytes, API frame, 118
dedicated gateways, 195
diagnostic commands, 288–289
Digi evaluation board, 7
Digi International, 1, 6, 263, 267
DigiKey, 6, 63
DIN pin, 15
DIO0...DIO12 Digital I/O pins, 15, 89
direct actuation, 171–187
    base station for, 177–180
    code for, 180–187
    coordinator for, 173
    parts for, 172
    routers for, 174–177
direct I/O, 88–89
direct phenomena, 86
distal phenomena, 87
distance, sensors for, 86
doorbell, 67–84
    breadboard for, connecting, 68
    breakout board for, connecting, 68–69
    button input for, 72–73, 77, 81
    buzzer output for, 73, 78, 83
    coordinator for, 68
    feedback for, 80–83
    nap doorbell, 83
    parts for, 67
    programs for, 77–80, 81–83
    router for, 68

large numbers, representing as bytes, 135
layers, ZigBee protocol, 25–26, 30, 236–240
LEDs, testing for power, association, or signal
        using, 79
length bytes, 115, 118
libraries for API, 141
        JPachube, 249
        NewSoftSerial, 225, 228
        xbee api, 250
light
        doorbell feedback using, 81
        romantic lighting sensor, 93–109
        sensors for, 86
LilyPad XBee, 269
link keys, 242
Linux
        adapter port, determining, 40
        terminal program for, 20, 40
        X-CTU program with, 33, 36
Linux:downloading Processing IDE, 150
local interactions, 112
logic-level serial (see TTL serial protocol)
LS Research, 267

M

MAC (media access controller) layer, 236
MAC address, 207
Macintosh
        adapter port, determining, 40
        configuring XBee radio, 43–46
        downloading Processing IDE, 150
        serial port for, 60
        terminal program for, 18, 19, 20
magnetic fields, sensor for, 86
MAKE: magazine, 6
Maker SHED, 6, 62
many-to-one routing, 240
Max/MSP library, 141
media access controller (MAC) layer, 236
Mega, Arduino, 59
mesh networking, 2, 26, 28, 236
messages stored for sleeping devices, 161–162
Microchip components, 267
microcontroller, 57
        (see also Arduino)
        external, 58
        not using, 88
microphone sensor, 86, 87
Mini, Arduino, 59

mobile data connections, 194
motion sensor, 86
Mouser, 63
multimeter, 13, 79, 104, 147, 159, 176, 226

N

nap doorbell, 83
network keys, 242
Network layer, 26, 236
networking commands, 278–281
networks, 27
        (see also wireless sensor networks; ZigBee
        network)
        connectivity between (see gateways)
        protocols for, 189
New Micros XBee Dongle, 9
NewSoftSerial library, 225, 228

O

ON pin, 15

P

Pachube site, 245–257
        account for, signing up, 246
        API key for, 248
        program for, 249–256
        registering a feed, 246
        troubleshooting sensor network using, 256
packet sniffers, 267
pair network, 27
PAN addresses, 29, 216
Panasonic, 267
parent device, messages stored by, 161–162
        (see also coordinator device; router device)
photocell, 86, 87
PHP code, running on XIG, 220–221
PHY (physical) layer, 25, 236
picocom program, 20, 40
pin configurations, 15, 89
ports, for adapter, 40
position, sensors for, 86
potentiometer, 86
power (VCC) pin, 15
pressure sensor, 86
PRO version, XBee radio, 2
Processing & Java library, 141
Processing IDE, 150–152, 262
profiles, application, 237–238

About the Author

Robert Faludi is an NYU professor, SVA professor, and an expert consultant on commercial projects, including large-scale home energy monitoring. His work has appeared in *The New York Times*, on CNET, on *Good Morning America*, and elsewhere. He is a co-creator of the LilyPad XBee wearable radios, and of Botanicalls, a system that allows thirsty plants to place phone calls for human help.

Colophon

The animals on the cover of *Building Wireless Sensor Networks* are dachshunds. The short-legged, elongated dogs were first bred for hunting in the 17th century in Germany; in fact, the name literally means "badger dog." Dachshunds are officially classified as members of the hound family in the United States, though there are some that argue that many varieties, especially wire-haired types, look and behave more like members of the terrier group. The World Canine Organization, which boasts 86 member countries, takes a middle road and specifies a separate group for dachshunds apart from both terriers and other scent hounds.

Further disagreement has arisen involving the official varieties of the dachshund breed. The World Canine Organization defines three sizes: standard, miniature, and rabbit. The American Kennel Club, on the other hand, recognizes only the standard and miniature sizes, arguing that so-called "rabbit" dachshunds are just comparatively smaller miniature varieties. Three coat types are universally recognized, however: smooth-haired, long-haired, and wire-haired.

Though dachshunds are popular pets in the United States, the dogs are perhaps most strongly prized in and associated with Germany. While dachshunds there are generally called *Dackel*, worthy specimens that are able to pass blood-tracking tests earn the moniker *Teckel* and are held in higher regard. Furthermore, the animal is so popular that a dachshund, named Waldi, was chosen as the official mascot of the 1972 Summer Olympics in Munich.

The cover image is from *Lydekker's Royal History*. The cover font is Adobe ITC Garamond. The text font is Linotype Birka; the heading font is Adobe Myriad Condensed; and the code font is LucasFont's TheSansMonoCondensed.

The connection diagrams in this book were created with Fritzing, an open source tool for documenting, sharing, teaching, and designing interactive electronic projects. For more information, see their website: *http://fritzing.org*.

Get even more for your money.

Join the O'Reilly Community, and register the O'Reilly books you own. It's free, and you'll get:

- $4.99 ebook upgrade offer
- 40% upgrade offer on O'Reilly print books
- Membership discounts on books and events
- Free lifetime updates to ebooks and videos
- Multiple ebook formats, DRM FREE
- Participation in the O'Reilly community
- Newsletters
- Account management
- 100% Satisfaction Guarantee

Signing up is easy:

1. **Go to: oreilly.com/go/register**
2. **Create an O'Reilly login.**
3. **Provide your address.**
4. **Register your books.**

Note: English-language books only

To order books online:

oreilly.com/store

For questions about products or an order:

orders@oreilly.com

To sign up to get topic-specific email announcements and/or news about upcoming books, conferences, special offers, and new technologies:

elists@oreilly.com

For technical questions about book content:

booktech@oreilly.com

To submit new book proposals to our editors:

proposals@oreilly.com

O'Reilly books are available in multiple DRM-free ebook formats. For more information:

oreilly.com/ebooks

Spreading the knowledge of innovators   oreilly.com

Have it your way.

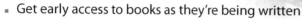

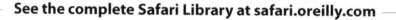

CPSIA information can be obtained
at www.ICGtesting.com
Printed in the USA
BVOW11s1800100616

451553BV00003B/6/P